Introduction

by Michael Horsley

Background - Lanchester is a mainly rural parish and, in common with other areas, the wildlife found there has changed and continues to change.

This wildlife audit is a snapshot based on historical and current records. It documents what is known of the wildlife of the parish with some indication of the changes over time. This brings the information together in a single document to illustrate to the people who live here the diversity around them and to provide input for future projects to protect and enhance this wildlife.

Various sites of interest for wildlife are mentioned in this document. Whilst there is public access to many, some of these are on private land and no right of access can be assumed.

Who has contributed - Many people have been involved in various ways, from providing the records of local wildlife sightings in the parish over many years and photographs of this wildlife, to the compilation of the sections of this audit. The sections have all been compiled by volunteers. The organisations listed at the back have also contributed by allowing their data to be used.

Summary - Because of its geography stretching from the Pennine foothills in the west to the gentler farmland in the east, the parish has a wide variety of habitats and supports a wide range of species from plants through to birds, invertebrates and amphibians. Although many are common, some are rare and declining and need help if they are to survive here and maintain the diversity for the enjoyment and benefit of future generations.

Method - Where available specific records from many sources have been brought together by the volunteers involved and analysed to provide a picture of wildlife in the parish. Elsewhere the knowledge of local experts has been drawn on.

Publication - This document has been made available in printed form as a book and can be viewed on-line and downloaded from the Lanchester Parish Council website at: www.lanchesterparish.info.

Lanchester Parish church: photo, Darin Smith

Small Tortoiseshell Butterfly, photo Darin Smith

Small Copper Butterfly, photo Darin Smith

Otter, photo Darin Smith

Peacock Butterfly, photo Darin Smith

Badger, photo Sue Charlton

The Locality Map Project - The compilation and publication of this wildlife audit has been part of a wider project to implement the "Lanchester Locality Map". This is a document compiled by Lanchester Parish Council, Lanchester Partnership and representatives of the farming, wildlife and environment groups of the parish to address rural issues. Funding was obtained for a Project Officer, Sue Charlton, to co-ordinate delivery of the Locality Map Action Plan. Whilst this audit has been in preparation other actions have included:

• A programme of educational visits and activities to connect local children with the environment and where food comes from;

• A monthly practical conservation volunteer team which has undertaken woodland and wetland management, invasive species control and tree planting;

• Support to farmers in securing grants such as English Woodland Grant Scheme and Higher Level Stewardship;

• A skills audit which has informed a programme of training that has included accredited training for young farm workers such as first aid, chainsaw, shearing and hedge laying;

• Training to support the wildlife and heritage audits including botanical survey, use of the Map Mate computer program and archival research;

• Botanical surveys of selected areas in the parish;

• A heritage audit resulting in the publication of 7 walk leaflets through the surrounding landscape that highlight the local heritage.

The Future - It is hoped that it will be possible under the auspices of the Locality Map Project and based on this wildlife audit to identify other sites in the parish which may contain interesting flora and fauna and to conduct survey work on those sites. It would also be beneficial to revisit sites for old records of rare plants to check whether the plants are still there.

By describing the range of local wildlife this audit can help people to develop a more informed knowledge of and interest in the natural environment around them. Anyone can already record interesting old trees on-line through the Woodland Trust website: www.ancient-tree-hunt.org.uk.

As a result of new and draft national planning legislation further local planning documents are in preparation by Durham County Council. The information contained in this audit can provide input to this process.

Landscape History

by John Gall

The pre-Roman occupation of the parish is not well recorded at present, but there have been finds of 'cup and ring' stones and flint tools in the valley and surrounding hills, indicating that there has been at least 4000 years of hunting and farming around Lanchester.

When the Roman legions marched north, leading to the creation of a fort and associated civil settlement of Longovicium, this would have had a considerable effect on the lands around the valley. Five hundred horse, (even small ones), the soldiers and their supporting community would have needed to be fed and watered and this could have involved the clearance of woodland and the cultivation of the best of the drained lands. Dere Street, the stone fort with its associated buildings and vicus would have required major quarrying and there are indications in the parish of extensive iron working which would have necessitated the felling of large areas of woodland in order to provide charcoal.

We know little of the period from the leaving of the Roman Garrison until the arrival of the Normans. There may have been a major estate centred on Lanchester but it does appear that native woodlands would have re-established themselves over much of the Roman farmlands.

With the arrival of Norman overlords activities were centred on the Bishopric and Church at Durham. During this time we have evidence of man's effect on the parish with further clearance of woodlands as well as the establishment of the present village around the church. There was some mediaeval iron working and the outcrop coal seams were worked in a small way. From late mediaeval times until the early 18th century the parish appears to have been a 'backwater' though much of the major woodland seems to have been felled and was replaced by coppice woodland for the production of hazel corf rods for the coal industry.

Lanchester Roman Fort, photo Darin Smith

The lands on the hill tops and to the west of the parish consisted of open fell for the grazing of animals owned by the freeholders and other residents. These lands were gradually enclosed with a major Enclosure Act sweeping up the last of the great open fells, in the late 1700s. Many miles of new fences, stone walls, roads and farm steadings were created or extended during this period. New plantations were developed and the old rough fell lands were cleared and brought into field agriculture. This 'enclosure' shaped the landscape into the form we know today.

The period from the Industrial Revolution of the late 1700s until the 20th century saw major changes with the creation of new communities associated with the exploitation of productive coal seams to the east of the parish. Railways, quarries, brickworks, iron and coke works brought much prosperity which peaked in the early 1900s and then decayed towards the later years of the 20th century.

A desperate shortage of coal in the years from the hard winter of 1947 led to most of the parish's outcrop coal being extracted by opencast or strip mining. In the Lanchester area of the coalfield, up to 50% of the land was ripped up, the coal extracted and the landform re-instated in a very average way, destroying much of the evidence of 4000 years of occupation, together with a loss of many habitats of plant and animal life.

The 21st century sees the parish settling down and recovering from the exploitation of coal and its associated industrial processes. New strategies at a national and European level are encouraging all areas of the community to take greater care of the environment and land we live in. We need to remember that we live in a much damaged landscape and great care is needed to re-instate the quality and diversity of environment of the entire parish.

Bumblebee and Bluebell, photo Darin Smith

Habitats

by Terry Coult

Heather, photo Darin Smith

Durham County Council's Landscape Character Assessment places Lanchester mostly in the West Durham Coalfield but rising and stretching far enough west to reach the North Pennines and coincidentally the Area of Outstanding Natural Beauty. The long history of land management in the parish is reflected broadly in the existing modern day habitats. After enclosure, ease of working and drainage along with accessibility produced a pattern of land use which is predominantly arable and pastoral. Some of the open heath survived subsequent tree planting, clearance and grazing towards the west of the parish. The more recent open cast mining made little change to this pattern. In summary land use changes from agricultural in the east of the parish through pastoral to moorland in the west.

Land use history, changes in altitude and a varied topography coupled with the fact that some areas of the parish have not suffered too severely from agricultural intensification means that the parish has a wide range of habitats, including heathland, woodlands, unimproved grasslands, hedgerows, mature trees, rivers, streams and ponds. Such a variety of mature habitats allows the parish to support a diverse range of wildlife.

The parish has suffered in the past from large scale industrial activities such as opencast and deep coal mining and where restoration after coaling has taken place landscapes are often denuded of character and habitats, with subsequent loss of species. The parish does have some brownfield land, land which was previously used for industry but through time has reverted to a more natural state, the railway walks and the Malton Nature Reserve are the best examples of this. Often brownfield land supports many more species than the farmed land around it, with the Malton Nature Reserve possibly being the best example. Land management fashions within the parish are still changing and the areas of heathland shown on the map in the west of the parish are still reducing in size as agricultural intensification continues. In the last few years much of the remaining upper Browney Valley heathland has been ploughed and fertilised with deleterious effect on the resident upland wildlife.

There are no measurable figures for habitats within the parish. It would be useful to say that there are so many miles of hedgerow and so many hectares of woodland and heathland but those figures are not calculated. Perhaps the best way to understand what is present is to just look at the "Landuse Map". It is easy to see that most of the parish is pasture land and that most of what little woodland there is, is conifer plantation, there being very little deciduous woodland, with only three recorded patches of ancient woodland in the parish, Deanery Wood at Ornsby Hill, Loves Wood and part of the river bank woodlands at Malton. It is likely that there will be smaller patches of unrecorded ancient woodland scattered around the parish and there are certainly many copses and stands of mature deciduous trees which are not big enough to register on the map, including hedgerow trees. Like many parts of the county and the country these mature trees are not matched in number by younger replacement trees and in years to come there is likely to be a dearth of trees around the parish. The parish has some very rich road verges in the west, verges which act as refuges for plants and animals which once would have occupied the surrounding farmed land but without management these verges will eventually lose their wildlife value.

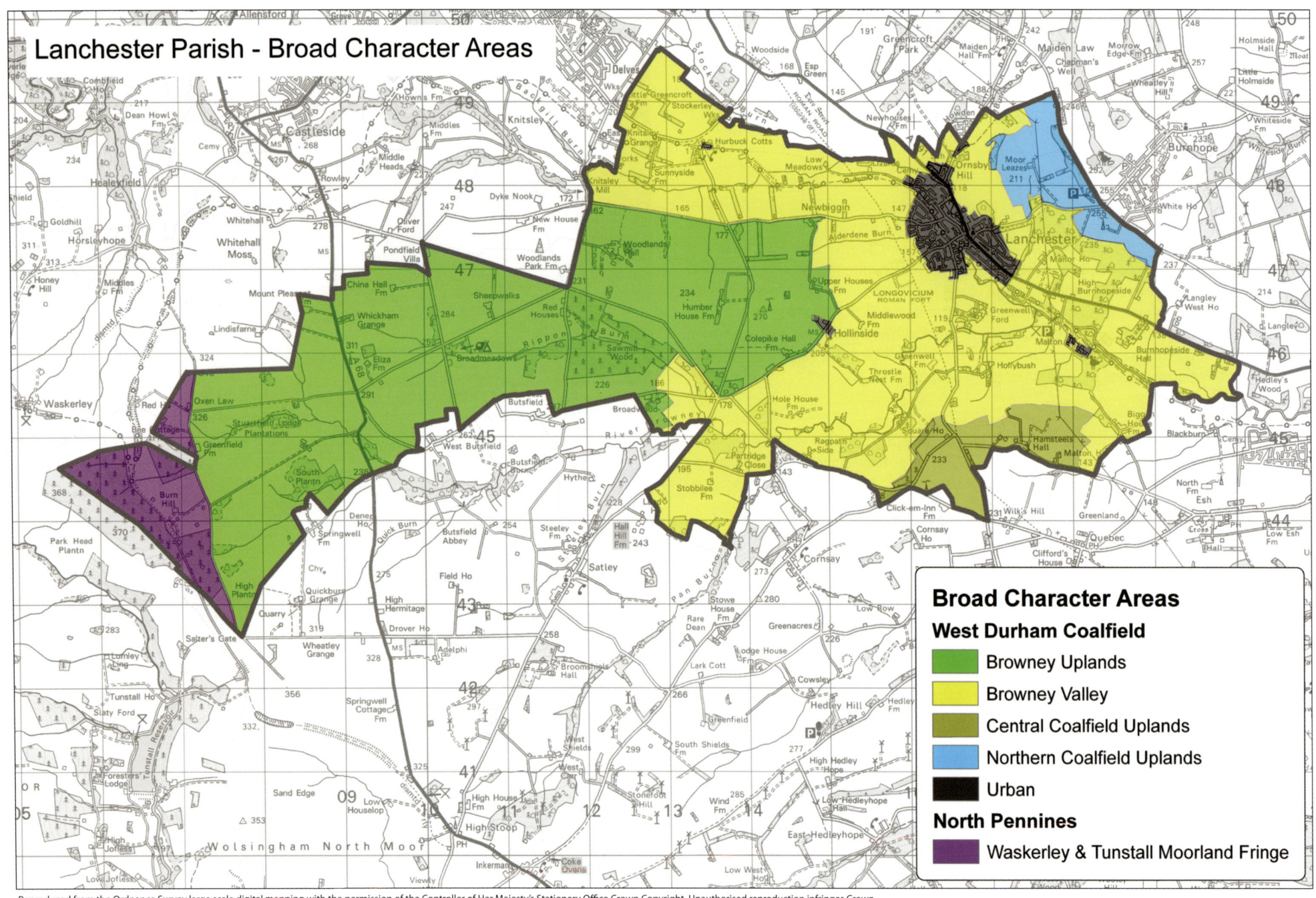

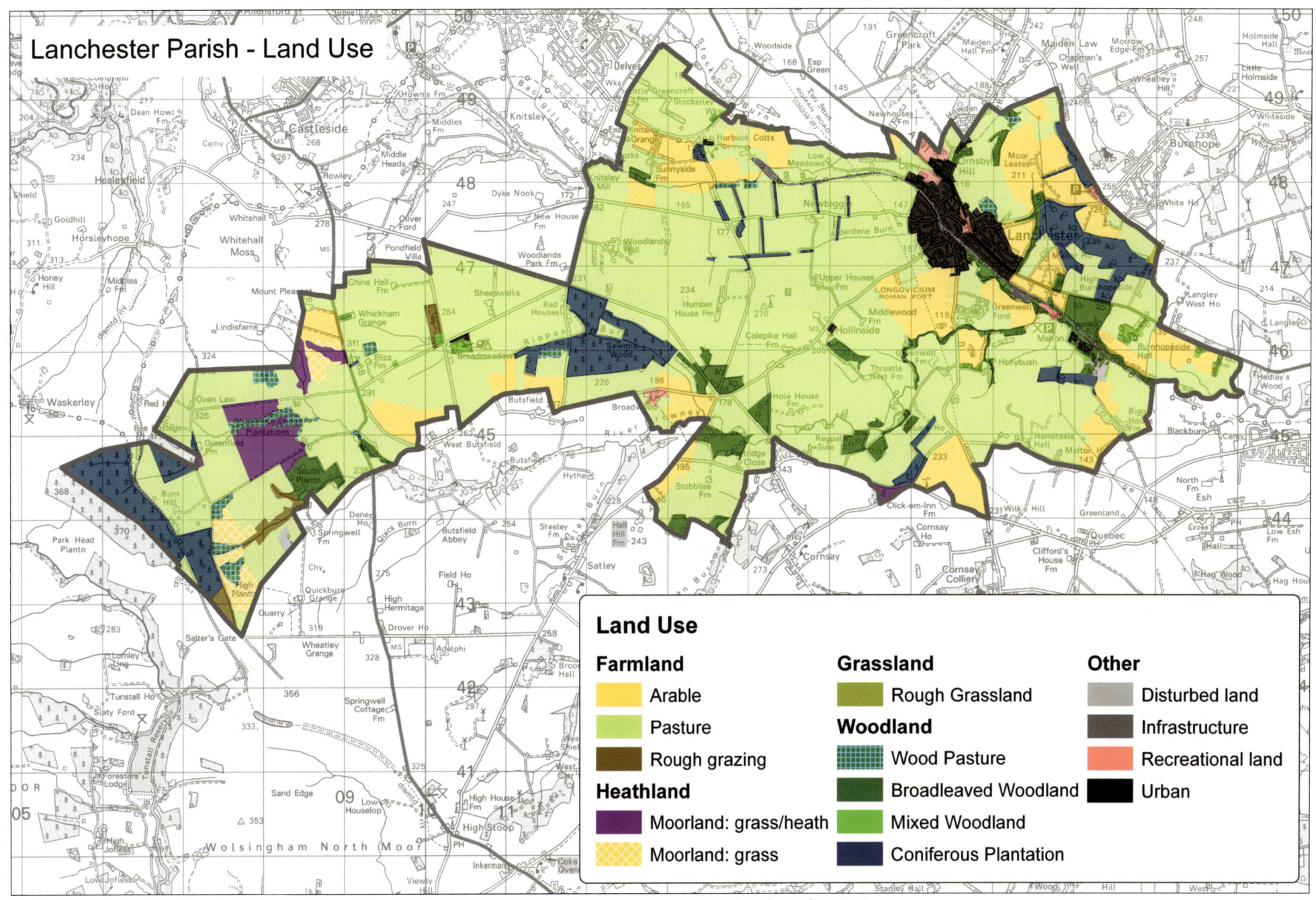

Malton, photo Sue Charlton

Longburn Ford, photo Durham Wildlife Trust

In terms of protected habitats Durham Wildlife Trust has nature reserves at Malton and Longburn Ford at the extreme ends of the parish and at Ragpathside and Burnhope just outside the parish. The Woodland Trust has woodland reserves at Black Plantation near Satley and at Dora's wood in Lanchester itself. The rest of the parish is very much in the hands of those who manage the land.

It is difficult to predict how land use will change in the future. Farming practices are very much driven by the latest fashions in grant aid and there is a very powerful incentive to sacrifice traditional land use and management in favour of diversification into tourism, with consequent increased recreational use of the land. Such changes have consequences for wildlife. Comparison of historical records with current ones show a decline in species and habitats across the parish and it is likely that without a great deal of investment the decline will continue.

Ragpath Heath, photo Durham Wildlife Trust

Wildflowers and Agricultural land, photo Darin Smith

Malton Pond, photo Darin Smith

Plants

by Angela Horsley

For the purposes of this audit some 7000 plant records have been assembled and studied. These have been supplied by:
- John Durkin, Durham County Recorder for the Botanical Society of the British Isles (BSBI)
- Durham Biodiversity Data Service (DBDS)
- Lanchester Wildlife Group (LWG)
- Durham Rare Plant Register published by the BSBI
- Durham Biodiversity Partnership
- Durham County Council.

Most of these records fall in the period 1970-2010 but there were a number of older records from 1945, 1961 and a handful from the mid 1800s.

The records have been made over time by different people and organisations for different purposes and from different features of interest such as Lanchester Valley Walk, nature reserves and Local Wildlife Sites. They are not comprehensive and individual records have not generally been re-visited so it is not known whether plants recorded some years ago are still there. So whilst this section of the audit considers what plants of interest have been recorded in the parish, there is no guarantee that any particular plant still exists where it was recorded.

Special Sites
There are six sites in the parish designated as Local Wildlife Sites (LWSs) by Durham County Council as they contain habitats or species with a special value for biodiversity.

Loves Wood and Malton Nature Reserve NZ180457
This local nature reserve has mixed habitats of woodland, ponds, neutral grassland and scrub. These support a diverse range of plants.

Loves Wood has a central block of mature Oak (*Quercus sp.*) woodland with occasional Birch (*Betula sp.*) in the canopy and an understorey of coppiced Hazel (*Corylus avellana*) and Holly (*Ilex aquifolium*). The ground flora contains Wood Sorrel (*Oxalis acetosella*). The core of oak woodland is surrounded by blocks of conifers – Spruce (*Picea sp*), Pine (*Pinus sp*) and Larch (*Larix sp*) which contain patches of Oak and Ash (*Fraxinus excelsior*). The Larch plantation has a good ground flora.

Frog Orchid, photo Michael Horsley Butterfly Orchid, photo Terry Coult

Malton Nature Reserve has been developed on a reclaimed colliery site with a wide variety of habitats including woodland, ponds, species rich grasslands and scrub showing good examples of succession on colliery shales. There are two ponds with a range of wetland plants, including Bog Bean (*Menyanthes trifoliata*) and adjacent willow carr has several uncommon plants such as Skullcap (*Scutellaria galericulata*). The scarcer Lesser Skullcap (*Scutellaria minor*) has also been recorded on the reserve. There are several small species rich meadows containing Bird's-foot Trefoil (*Lotus corniculatus*), Devil's-bit Scabious (*Succisa pratensis*) and Tormentil (*Potentilla erecta*). An old hedgerow and area of oak woodland contain plants indicative of ancient woodland such as Moschatel (*Adoxa moschatellina*) and Dog's Mercury (*Mercurialis perennis*). Several garden escapes can be found particularly on the southeastern side of the site such as Jacob's Ladder (*Polemonium caeruleum*) and Yellow Archangel (*Lamiastrum galeobdolon*).

Greenwell Ford Meadow NZ166464
When this LWS was surveyed in 1991 the central section of the meadow contained a pond derived from a former oxbow section of the River Browney. This is of great interest as this type of habitat is now scarce in the Browney Valley.

The pond area was surrounded by Alder (*Alnus glutinosa*), Willow (*Salix sp*), Sessile Oak (*Quercus petraea*), Rowan (*Sorbus aucuparia*), Beech (*Fagus sylvatica*), Large-leaved Lime (*Tilia platyphyllos*) and some conifers. The pond itself contained many plants including Yellow Iris (*Iris pseudacorus*), Marsh Marigold (*Caltha palustris*) and Water-crowfoot (*Ranunculus sp*).

The central part of the meadow was uncut and ungrazed and was damp in places with a good variety of herbs and grasses including Yellow Rattle (*Rhinanthus minor*), Cuckooflower (*Cardamine pratensis*), Crosswort (*Galium cruciata*) and Great Burnet (*Sanguisorba officinalis*).

Hurbuck Triangle NZ143481
This LWS surveyed in July 2007 comprises a particularly species rich stretch of former railway line which forms part of the Lanchester Valley Walk together with a small triangle of wet grassland adjacent to the railway on the northern side.

The habitat varies along the railway and includes patches of herb rich neutral grassland, acid grassland and areas of wet grassland. Greater Butterfly Orchids (*Platanthera chlorantha*) have been recorded on this site in the past, both on the railway line and within the triangle but none were seen during survey in 2007. There are many species of interest along the railway line including:

Common name	Specific name
Bladder Campion	*Silene vulgaris*
Burnet Saxifrage	*Pimpinella saxifraga*
Common Bird's-foot Trefoil	*Lotus corniculatus*
Common Knapweed	*Centaurea nigra*
Common Spotted Orchid	*Dactylorhiza fuchsii*
Cowslip	*Primula veris*
Eyebright	*Euphrasia sp*
Field Scabious	*Knautia arvensis*
Goatsbeard	*Tragopogon pratensis*
Great Burnet	*Sanguisorba officinalis*
Kidney Vetch	*Anthyllis vulneraria*
Lady's Bedstraw	*Galium verum*
Oxeye Daisy	*Leucanthemum vulgare*
Pale Lady's Mantle	*Alchemilla xanthochlora*
Quaking Grass	*Briza media*
Slender St John's Wort	*Hypericum pulchrum*
Smooth Lady's Mantle	*Alchemilla glabra*
Tormentil	*Potentilla erecta*
Zigzag Clover	*Trifolium medium*

Towards the eastern end of the site, there are raised banks with thin soils. These are more acidic in nature with Catsear (*Hypochaeris radicata*), Harebell (*Campanula rotundifolia*), Mouse-ear Hawkweed (*Pilosella officinarum*), Sheep's Fescue (*Festuca ovina*), Heath Bedstraw (*Galium saxatile*) and Betony (*Stachys officinalis*).

Cowslip, photo Darin Smith

Mouse-ear-hawkweed, photo Darin Smith

Common Dog Violet, photo Darin Smith

Harebell, photo Darin Smith

Bog Bean, photo Terry Coult

Marsh Marigold, photo Darin Smith

Common Spotted Orchid, photo Darin Smith

Cotton Grass, photo Darin Smith

At the western end of the site, the railway sides slope steeply downwards and the triangle is located at the base of one of these steep slopes on the northern side. It is mostly dominated by Meadowsweet (*Filipendula ulmaria*), with Common Valerian (*Valeriana officinalis*), Wild Angelica (*Angelica sylvestris*) and Rose-bay Willowherb (*Chamerion angustifolium*).

The triangle appears to have lost a number of the species that were previously recorded in it and it requires more management with a heavier grazing level to hopefully restore it.

Stuartfield Moor NZ086447
This LWS lies at the western end of the parish and comprises heathland and woodland habitats. It has three sections and was surveyed in 1992 and 1994.

North Plantation and Stuartfield Lodge is an area of mid-altitude heathland which is the only area of this habitat in the parish. The heathland is dominated by heather and bilberry. A number of uncommon plant species are found along the southern edges of the site including Adder's tongue (*Ophioglossum vulgatum*), Moonwort (*Botrychium lunaria*) and Petty Whin (*Genista anglica*).

South Plantation and Woodburn Plantation – the woodland is dominated by Birch (*Betula pubescens* and *Betula pendula*) with some mature Sessile Oak (*Quercus petraea*) and Rowan (*Sorbus aucuparia*) and a small amount of Spruce (*Picea sp.*) and Larch (*Larix sp.*). The ground flora consists of grasses and Bracken (*Pteridium aquilinum*) with patches of Wood-sorrel (*Oxalis acetosella*) and Foxglove (*Digitalis purpurea*) scattered throughout. Alder (*Alnus glutinosa*) lines the stream with Eared Willow (*Salix aurita*) by a small pond. Wetter patches contain Sharp-flowered Rush (*Juncus acutiflorus*) and Marsh Violet (*Viola palustris*).

High Plantation is an area of mid-altitude heather moor and birch plantation grazed by sheep. The ground layer is Heather (*Calluna vulgaris*) and Bilberry (*Vaccinium myrtillus*) with occasional patches of Cross-leaved Heath (*Erica tetralix*) and Crowberry (*Empetrum nigrum*) with Sphagnum moss and Hare's-tail Cottongrass (*Eriophorum vaginatum*) in wetter areas. Birch (*Betula pubescens* and *B. pendula*) forms open woodland in part of the area while in other parts Birch are scattered together with occasional Scots Pine (*Pinus sylvestris*).

Burnhill Junction and Longburn Ford NZ070444
This LWS comprises several different habitats which are important for the Small Pearl-bordered Fritillary butterfly, that is, acid grassland, meadow and young tree planting. It was surveyed in June 2004.

Burnhill Junction is a section of the Waskerley Way and contains Marsh Violet (*Viola palustris*) in a gutter to the west of the track on which the larvae of the butterfly feed.

Some of the meadows at the apex of the junction have had Violets introduced and appropriate shelter planted in the hope that the butterfly will also establish there. To the east of the junction the field bordering the Browney is quite sheltered at the bottom of the valley and has been planted with enclosures of Alder (*Alnus glutinosa*), Juniper (*Juniperus communis*), Rowan (*Sorbus aucuparia*) and Silver Birch (*Betula pendula*).

Longburn Ford Quarry is also managed for the same butterfly. There are many plants of interest.

Low-lying area to west side of road opposite quarry
Ragged Robin	*Lynchis flos-cuculi*
Valerian	*Valeriana officinalis*
Marsh Violet	*Viola palustris*

On the roadside
Bell Heather	*Erica cinerea*
Bilberry	*Vaccinium myrtillus*
Bird's-foot Trefoil	*Lotus corniculatus*
Dog Violet	*Viola canina*
Oxeye Daisy	*Leucanthemum vulgare*
Pignut	*Conopodium majus*
Quaking Grass	*Briza media*
Wavy Hair Grass	*Deschampsia flexuosa*

Field to west of road
Hare's-tail Cottongrass	*Eriophorum vaginatum*
Heath-spotted Orchid	*Dactylorhiza maculata subsp. ericetorum*
Star Sedge	*Carex echinata*

Black Plantation NZ137450
This is now owned by the Woodland Trust and managed as a woodland reserve. It was surveyed in August 1990.

The western section is a wet acid birch wood with a very sparse understorey. The ground flora is mostly grasses (*Holcus mollis* and *Holcus lanatus*) and ferns (*Dryopteris filix-mas*) with Sphagnum moss in the depressions.

The eastern section of the wood is much drier, mainly Birch with a small amount of Sessile Oak (*Quercus petraea*) and a few Beech (*Fagus sylvatica*). The understorey is better developed with a variety of shrub species including Holly (*Ilex aquifolium*), Hazel (*Corylus avellana*), Guelder-rose (*Viburnum opulus*) and Bird Cherry (*Prunus padus*). The herb layer includes Dog's Mercury (*Mercurialis perennis*), Heather (*Calluna vulgaris*) and Bilberry (*Vaccinium myrtillus*).

The woodland also has two large clearings which were surveyed in June 2007 and a total of ten sedges were recorded.

Clearing 1 lies to the west NZ135449 and is kept clear because of overhead pylons. It is mostly an acid grassland community including:

Carnation Sedge	*Carex panacea*
Devil's-bit Scabious	*Succisa pratensis*
Field Woodrush	*Luzula campestris*
Green-ribbed Sedge	*Carex binervis*
Hairy Woodrush	*Luzula pilosa*
Heath Bedstraw	*Galium saxatile*
Heath Woodrush	*Luzula multiflora*
Marsh Violet	*Viola palustris*
Pale Sedge	*Carex pallescens*
Pignut	*Conopodium majus*
Pill Sedge	*Carex pilulifera*
Tormentil	*Potentilla erecta*
Wavy Hair Grass	*Deschampsia flexuosa*

Clearing 2 lies to the east NZ138449 and is mostly fen habitat. It has many species of interest including eight sedges:

Bay Willow	*Salix pentandra*
Betony	*Stachys officinalis*
Bog Stitchwort	*Stellaria uliginosa*
Bottle Sedge	*Carex rostrata*
Carnation Sedge	*Carex panacea*
Common Sedge	*Carex nigra*
Common Spotted Orchid	*Dactylorhiza fuchsii*
Common Valerian	*Valeriana officinalis*
Cuckooflower	*Cardamine pratensis*
Devil's-bit Scabious	*Succisa pratensis*
Fen Bedstraw	*Galium uliginosum*
Glaucous Sedge	*Carex flacca*
Great Burnet	*Sanguisorba officinalis*
Greater Tussock Sedge	*Carex paniculata*
Marsh Bedstraw	*Galium palustre*
Marsh Valerian	*Valeriana dioica*
Pale Sedge	*Carex pallescens*
Purple Moor Grass	*Molinia caerulea*
Ragged Robin	*Lychnis flos-cuculi*

Sanicle	*Sanicula europaea*
Slender St John's Wort	*Hypericum pulchrum*
Star Sedge	*Carex echinata*
Wood Sedge	*Carex sylvatica*

Road Verges

Five areas of road verge in the parish have been identified as still remaining species-rich and containing many interesting plants.

Salter's Gate to A68 NZ0774542630 – NZ0812843405

This was surveyed in July 2005. It is a wide verge characteristic of this area. Of the 52 species of plants recorded in that stretch of verge, several of interest were:

Bilberry	*Vaccinium myrtillus*
Goatsbeard	*Tragopogon pratensis*
Heather	*Calluna vulgaris*
Lady's Bedstraw	*Galium verum*
Melancholy Thistle	*Cirsium heterophyllum*
Oxeye Daisy	*Leucanthemum vulgare*
Ragged Robin	*Lychnis flos-cuculi*
Sneezewort	*Achillea ptarmica*
Tormentil	*Potentilla erecta*

Oxen Law to Salter's Gate NZ0736443991

46 species were recorded including:

Adder's Tongue	*Ophioglossum vulgatum*
Bilberry	*Vaccinium myrtillus*
Bird's-foot Trefoil	*Lotus corniculatus*
Harebell	*Campanula rotundifolia*
Heather	*Calluna vulgaris*
Quaking Grass	*Briza media*
Tormentil	*Potentilla erecta*
Yellow Rattle	*Rhinanthus minor*

Green Lane, north of West Butsfield NZ1023346078

This was surveyed in July 2005 and 55 species were found including of interest:

Betony	*Stachys officinalis*
Sweet Cicely	*Myrrhis odorata*
Wood Cranesbill	*Geranium sylvaticum*

Devil's Bit Scabious, photo Sue CharltonRagged Robin, photo Darin Smith

Quaking Grass, photo Darin SmithAdder's Tongue, photo Darin Smith

Wood Cranesbill, photo Darin SmithBilberry, photo Darin Smith

Road between West Lane and A68 NZ0937644564 – NZ0997344839
This was surveyed in July 2005 and 51 species were found including:

Angelica	*Angelica sylvestris*
Betony	*Stachys officinalis*
Common Valerian	*Valeriana officinalis*
Honeysuckle	*Lonicera periclymenum*
Knapweed	*Centaurea nigra*
Sweet Cicely	*Myrrhis odorata*

West Lane – south facing side NZ0998744843
Again this was surveyed in July 2005 and 53 species were found including:

Bird's-foot Trefoil	*Lotus corniculatus*
Field Scabious	*Knautia arvensis*
Harebell	*Campanula rotundifolia*
Lady's Bedstraw	*Galium verum*
Sweet Cicely	*Myrrhis odorata*
Yellow Rattle	*Rhinanthus minor*

In conjunction with Durham Biodiversity Partnership further road verges in the west of the parish have been surveyed during 2011 and more species rich sections found. The report will not be available until after printing of this audit.

Bird's-foot Trefoil, photo Darin Smith

Marsh Cinquefoil, photo Terry Coult

Oxeye Daisy, photo Darin Smith

Eyebright, photo Terry Coult

Celandine, photo Terry Coult

Harts Tongue Fern, photo Terry Coult

Veteran and Notable Trees
The Woodland Trust has been encouraging the general public to record trees of note in their area. This can be done through www.ancient-tree-hunt.org.uk as individuals. As part of this recording the Durham Biodiversity Partnership has a project to train volunteers in tree identification and recording. The table below sets out the trees recorded so far in the parish although it can in no way be considered complete.

Common	Specific name	Status	Grid ref	Girth (m)	Accessibility
Ash	*Fraxinus excelsior*	Veteran	NZ16584730	3.60	Public – open access
Ash	*Fraxinus excelsior*	Veteran	NZ16674765	4.40	Public – footpath
Beech	*Fagus sylvatica*	Notable	NZ16614772	3.0	Private – garden
Beech	*Fagus sylvatica*	Notable	NZ18224612	3.0	Public – footpath
Beech	*Fagus sylvatica*	Notable	NZ18054615	3.15	Private – visible from public access
Beech	*Fagus sylvatica*	Notable	NZ18104613	3.35	Private – visible from public access
Beech	*Fagus sylvatica*	Veteran	NZ09394573	3.50	Public – roadside
Beech	*Fagus sylvatica*	Veteran	NZ16704760	3.90	Private – visible from public access
Beech	*Fagus sylvatica*	Veteran	NZ16604730	4.50	Public – open access
Beech	*Fagus sylvatica*	Veteran	NZ14514593	5.50	Private – garden
Oak	*Quercus sp*	Notable	NZ15884850	3.31	Private – visible from public access
Oak	*Quercus sp*	Veteran	NZ17714634	4.0	Public – open access

Common	Specific name	Status	Grid ref	Girth (m)	Accessibility
Oak	*Quercus sp*	Notable	NZ17524754	4.40	Private – visible from public access
Oak	*Quercus sp*	Veteran	NZ16754767	4.50	Public – footpath
Silver birch	*Betula sp*	Veteran	NZ17414708	2.10	Private
Sycamore	*Acer pseudoplatanus*	Veteran	NZ13174610	3.20	Public – roadside
Sycamore	*Acer pseudoplatanus*	Veteran	NZ06364472	3.50	Public - footpath
Sycamore	*Acer pseudoplatanus*	Veteran	NZ15094646	3.50	Public – roadside
Sycamore	*Acer pseudoplatanus*	Veteran	NZ15104647	4.25	Public – roadside
Sycamore	*Acer pseudoplatanus*	Veteran	NZ16014699	4.30	Public – roadside
Sycamore	*Acer pseudoplatanus*	Notable	NZ1664447450	2.83	Public – open access
Horse Chestnut	*Aesculus hippocastanum*	Veteran	NZ1666047475	3.32	Public – open access
Sycamore	*Acer pseudoplatanus*	Veteran	NZ1667147460	3.32	Public – open access
Horse Chestnut	*Aesculus hippocastanum*	Notable	NZ1667047445	2.87	Public – open access
Beech	*Fagus sylvatica*	Veteran	NZ1670547597	3.50 (estimate)	Private – visible from public access
Sessile Oak	*Quercus petraea*	Notable	NZ1672447630	3.50	Public – footpath
Beech	*Fagus sylvatica*	Veteran	NZ1673544891	3.80	Roadside
Beech	*Fagus sylvatica*	Veteran	NZ1751344780	3.88	Roadside
Beech	*Fagus sylvatica*	Veteran	NZ1860145032	3.75 (estimate)	Private
Sessile Oak	*Quercus petraea*	Notable	NZ1792545691	3.42	Private
Silver birch	*Betula sp*	Veteran	NZ1788545654	1.50	Private
Rowan	*Sorbus aucuparia*	Veteran	NZ1789045660	2.30	Private
Sessile Oak	*Quercus petraea*	Notable	NZ1791545632	2.80	Private
Sessile Oak	*Quercus petraea*	Notable	NZ1792545627	2.80	Private
Sessile Oak	*Quercus petraea*	Notable	NZ1795945609	2.72	Private
Silver birch	*Betula sp*	Ancient	NZ1794545637	1.95	Private
Juniper	*Juniperus communis*	Ancient	NZ1440444695	4.5 circ.	Public – open access
Sessile Oak	*Quercus petraea*	Notable	NZ1398044538	3.0 (estimate)	Roadside
Sessile Oak	*Quercus petraea*	Notable	NZ1397744516	2.75 (estimate)	Private – visible from public access
Ash	*Fraxinus excelsior*	Notable	NZ1397944524	2.75 (estimate)	Private – visible from public access
Beech	*Fagus sylvatica*	Ancient	NZ0882244497	4.90	Private – visible from public access
Beech	*Fagus sylvatica*	Ancient	NZ0878644485	5.40	Private – visible from public access
Sycamore	*Acer pseudoplatanus*	Ancient	NZ1299045526	4.50 (estimate)	Private – visible from public access

Oak leaf, photo Thinkstock

Rowan, photo Darin Smith

Juniper, photo Thinkstock

Horse chestnut, photo Thinkstock

Silver birch, photo Thinkstock

Beech, photo Thinkstock

Invasive Plants

The records show two very invasive plants in the parish. The first is Himalayan Balsam (*Impatiens glandulifera*) which is recorded along the Lanchester Valley Walk and by the River Browney in the Malton area of the parish. It is visibly increasing year by year very rapidly in this area.

The second is Japanese Knotweed (*Fallopia japonica*) which is recorded in the Hurbuck Triangle but so far nowhere else in the parish.

There are also a few Rhododendrons (*Rhododendron ponticum*) recorded along the Lanchester Valley Walk near Malton but again they do not seem to be invasive in the parish.

Rare Plants

The Botanical Society of the British Isles (BSBI) has published a number of county Rare Plant Registers which have been prepared by their County Recorders which detail the rarest species in a county. The Durham Rare Plant Register, produced by Durham County Recorder John Durkin, was published in 2010 and contains details of both nationally rare species and locally rare and scarce species. The nationally rare species are those listed in the BSBI / IUCN (International Union for Conservation of Nature) list and are classified according to their level of rarity, for example, critically endangered, endangered, rare, vulnerable, scarce etc. A local species is rare if it has been recorded in 3 or less sites in the county and scarce if it has less than 15 sites in the county.

There are records of a number of Rare Plant Register species in the parish. The national status is shown first for each species followed by the county status:

Coeloglossum viride   Frog Orchid   Vulnerable   Declining
04 Jun 2003	NZ0742	Salter's Gate	Lanchester Wildlife Group (LWG)
15 Jun 2005	NZ149480	Lanchester Way	A & G Young
10 Jul 2006	NZ0742	Salter's Gate	LWG (13 spikes recorded)

Platanthera chlorantha   Greater Butterfly Orchid   Near threatened   Scarce
2002	NZ143481	Hurbuck Triangle	Stobbs J. et al
10 Jun 2005	NZ143481	Hurbuck Meadow	A & G Young
15 Jun 2005	NZ143480	Lanchester Way – north side of track	A & G Young

Chenopodium bonus-henricus   Good King Henry   Vulnerable   Declining
| 02 Jul 1974 | NZ1447 | Newbiggin Farm | Mrs M Burnip |

Helleborus foetidus   Stinking Hellebore   Scarce   Probably hortal (of garden origin)
| 15 Jun 2005 | NZ164473 | Lanchester Way – on wall opposite old station | A & G Young |

Genista anglica   Petty Whin   Near Threatened   Scarce
| 27 May 1978 | NZ085455 | North Plantation | M. Shaw |
| Feb 1994 | NZ0845 | Stuartfield Moor LWS | Valerie Standen |

Viola canina   Heath Dog Violet   Near Threatened   Scarce
| Jun 2004 | NZ070444 | Burnhill Junction and Longburn Ford LWS | Stobbs J. et al |

Juniperus communis   Juniper   UKBAP   Durham BAP
10 Feb 1993	NZ158445	Ragpath	G. Lawson
Jun 2004	NZ070444	Burnhill Junction and Longburn Ford LWS	Stobbs J. et al
29 Sep 2004	NZ063446	Waskerley Way	A & G Young

Polemonium caeruleum   Jacob's Ladder   Nationally rare   Scarce hortal

There are several records of this plant at Malton Nature Reserve. It is thought to have originated from gardens of the houses that used to be on this site.

1991	NZ182459	Malton Nature Reserve	Mr R. Boyce
1991	NZ183457	Malton Nature Reserve	Mr R. Boyce
1992	NZ1845	Malton Nature Reserve	Mr R. Boyce
30 Aug 2006	NZ183457	Malton CWS 1.29, east field	John Durkin
27 Jun 2008	NZ1845	Malton Nature Reserve	Cleveland Naturalists Field Club

Scutellaria minor   Lesser Skullcap   Locally scarce
| 10 Aug 2005 | NZ1845 | Malton Nature Reserve | A & G Young |

Tilia platyphyllos   Large-leaved Lime   Scarce   Scarce, planted
| Jun 1991 | NZ166464 | Greenwell Ford Meadow | Valerie Standen |

Birds

by Fiona Green and Gary Bell

Red Kite, photo Darin Smith

J.W. Fawcett published Birds of Durham in 1890 and commented that a total of 126 birds were breeding in the county. Birds which were thought common then included Ring Ouzel (*Turdus torquatus*), Reed Warbler (*Acrocephalus streperus*) and Nightjar (*Caprimulgus europaeus*). Any unusual sightings were invariably shot including a Bewick's Swan shot at Bearpark in 1843.

Despite almost 120 years of conservation work since this inventory was compiled there are 27 species of birds at risk in County Durham. However at the end of this section a list shows 124 species that might be seen in the parish and many of them have been recorded making it a significant area for ornithology.

The landscape of Lanchester Parish rises to upland heath in the west and falls to the River Browney and tributaries. Between these contrasting settings the parish offers a wide variety of habitats including: gardens, farmland, deciduous and coniferous woodland. There are no large bodies of water within the parish.

Records have been sourced from surveys provided by the following groups: Durham Bird Club, Lanchester Wildlife Group and Durham Biodiversity Data Service. Observers include Gen McPartland, Gary Bell, G.W. Heslop, Alan Jones, Tom Oliphant, John Olley and David Sowerbutts.

Almost 500 records have been compiled, dating between 1986 and 2011. They are not comprehensive and vary in detail with some missing six digit grid references. Nevertheless they provide a valuable overview of birdlife in the parish.

Species are mentioned once although they may occur at many of the observation points. Others may not have been mentioned but are included in the list.
NB. Durham Biodiversity Action Plan Priority Species are marked (PS).

Types of habitat
Recording has taken place at several main observation points, Lanchester village, Dora's Wood, Manor House, Malton Nature Reserve, Malton Picnic Area, Ornsby Hill, Broadwood, Salter's Gate and Stuartfield Lodge.

House Sparrow, photo Darin Smith

Goshawk, photo Darin Smith

Snipe, photo Joe Ridley

Wren, photo Darin Smith

Fieldfare, photo Darin Smith

Barn Owl, photo Joe Ridley

Skylark, photo Darin Smith

Lapwing, photo Darin Smith

Chiff Chaff, photo Darin Smith

Kestrel, photo Darin Smith

Wheatear, photo Darin Smith

The Lanchester Wildlife Group observations are particularly useful for gaining a perspective on garden birds in the parish. The railway line corridor allows birds such as Siskins (*Carduelis spinus*), which congregate in flocks, to visit adjacent gardens.

Garden Birds – Lanchester Village NZ1647

These include ubiquitous species such as Sparrow Hawk (*Accipitur nisus*), Starling (*Sturnus vulgaris*) and Robin (*Erithacus rubecula*). House Sparrow (PS) (*Passer domesticus*) has also been noted. Winter visitors include Waxwing (*Bombycilla garrulus*), Brambling (*Fringilla montifringilla*) and summer visitors include Swift (*Apus apus*) and Spotted Flycatcher (*Muscicapa striata*).

Other species observed between 1986 and 2009 include:

Common name	Specific name	Status in Lanchester Parish
Green Woodpecker	*Picus viridis*	Resident
Collared Dove	*Streptopelia decaocto*	Resident
Kestrel	*Falco tinnunculus*	Resident
Cuckoo	*Cuculus canorus*	Summer visitor
Wren	*Troglodytes troglodytes*	Resident

Woodland and Farmland Birds
Dora's Wood – Lanchester Village NZ168469

Dora's Wood (Woodland Trust) lies on the south east side of the village and was planted in 2000. The Smallhope Burn runs to the south and west of the wood. Resident species recorded there include: Dunnock (*Prunella modularis*), Great Spotted Woodpecker (*Dendrocopos major*), Nuthatch (*Sitta europaea*) and Greenfinch (*Carduelis chloris*). Summer visitors include Swallow (*Hirondo rustica*) and Blackcap (*Sylvia atricapilla*). During the winter visiting Fieldfare (*Turdus pilaris*), Redwing (*Trudus iliacus*) and flocks of Long-tailed Tit (*Aegithalos caudatus*) can be seen. Other sightings include:

Goldcrest	*Regulus regulus*	Resident
Kingfisher	*Alcedo atthis*	Resident
Snipe	*Gallinago gallinago*	Resident
Sand Martin	*Riparia riparia*	Summer visitor
Grasshopper Warbler	*Locustella naevia*	Summer visitor

Malton Picnic Area – Lanchester Village NZ1746

Malton is a hamlet on the south east side of the parish and is where the Smallhope Burn joins the River Browney. The habitat is mainly woodland but borders farmland. Birds recorded along the water courses include Grey Heron (*Ardea cinerea*), Dipper (*Cinclus cinclus*) and Grey Wagtail (*Motacilla cinerea*). Tawny Owl (*Strix aluco*) and Kestrel (*Falco tinnunculus*) are recorded nesting there. The diverse habitat at Malton draws a wide variety of less common birds including:

Goosander	*Mergus merganser*	Summer visitor
Cormorant	*Phalacrocorax carbo*	Winter visitor
Redpoll (sp.)	*Carduelis flammea*	Winter visitor
Grey Partridge	*Perdix perdix*	Resident
Peregrine Falcon (PS)	*Falco peregrinus*	Vagrant
Woodcock	*Scolopax rusticola*	Resident
Oystercatcher	*Haematopus ostralegus*	Resident
Tree Sparrow	*Passer montanus*	Resident
Treecreeper	*Certhia familiaris*	Resident
Willow Warbler	*Phylloscopus trochilus*	Summer visitor
Tree Pipit	*Anthus trivialis*	Summer visitor

Malton Nature Reserve – Lanchester Village NZ183458

The reserve is located south east of Malton hamlet on reclaimed land which was the site of Malton Colliery. Jay (*Garulos glandarius*) and Pheasant (*Phasianus colchicus*) are often seen there. The vegetation consists of scrub and woodland and supports many of the warblers including:

Grasshopper Warbler	*Locustella naevia*	Summer visitor
Whitethroat	*Sylvia communis*	Summer visitor
Chiffchaff	*Phylloscopus collybita*	Summer visitor
Willow Warbler	*Phylloscopus trochilus*	Summer visitor

Manor House – Lanchester Village NZ1747

This area consists of farmland bordered by woodland and lies north east of the village centre. Woodpigeon (*Columba palumbus*), Mistle Thrush (*Turdus viscivorus*) and Blackbird (*Turdus turdus*) have been recorded. The woodland consists of deciduous and coniferous plantations and is populated by owls including:

Long-eared Owl	*Asio otus*	Resident

Ornsby Hill – Lanchester Village NZ 167483

Ornsby Hill lies on the northern edge of Lanchester and the habitat includes woodland and farmland. Chaffinch (*Fringilla coelebs*), Sparrowhawk (*Accipitur nisus*), Coal Tit (*Periparus ater*), and Yellowhammer (*Emberiza citronella*) have been observed in this area. Also House Martin (*Delichon urbica*) are summer visitors. Less common sightings are listed below:

Barn Owl (PS)	*Tyto alba*	Resident
Red Kite	*Milvus milvus*	Vagrant
Garden Warbler	*Sylvia borin*	Summer visitor
Carrion Crow	*Corvus corone*	Resident

Red Grouse, photo Joe Ridley

Little Owl, photo Darin Smith

Bullfinch, photo Darin Smith

Great Crested Grebe, photo Darin Smith

Grey Partridge, photo Darin Smith

Stonechat, photo Darin Smith

Goldfinch, photo Joe Ridley

Song Thrush, photo Darin Smith

Marsh Tit, photo Darin Smith

Long Eared Owl, photo Darin Smith

Broadwood – NZ 1245

Broadwood is a small hamlet approximately two miles west of Lanchester. The land is undulating and crossed by the River Browney. Broadwood includes woodland and land which is farmed mostly for sheep and arable crops. Birds recorded in this area include Linnet (PS) (*Carduelis cannabila*), Skylark (PS) (*Alauda arvenis*) and Song Thrush (PS) (*Turdus philomelos*). More unusual sightings include:

Spotted Flycatcher (PS)	*Muscicapa striata*	Summer visitor
Tree Sparrow (PS)	*Passer montanus*	Resident

Salter's Gate – NZ 0743

Salter's Gate is an area of upland heath which is located on the west boundary of the parish. The heath provides a rich, relatively undisturbed, habitat for a wide variety of species many of them at risk. Residents include Common Starling (PS) (*Sturnus vulgaris*). Summer visitors include Cuckoo (*Cuculus canorus*) and Redstart (*Phoenicurus phoenicurus*). Twite (*Carduelis flavirostris*) and Northern Lapwing (PS) (*Vanellus vanellus*) have been recorded there in winter. Other species observed there include Merlin (PS) (*Falco columbarius*), Black Grouse (PS) (*Tetrao tetrix*), Snipe (PS) (*Galingago galinago*), Eurasian Curlew (PS) (*Numenius arquata*) and Common Redshank (PS) (*Tringa totanus*). Other sightings include:

Hen Harrier (PS)	*Circus cyaneus*	Summer and Winter visitor
Buzzard	*Buteo buteo*	Resident
Goshawk	*Accipitur gentilis*	Resident
Brambling	*Fringilla montifringilla*	Winter visitor

List of the Birds of the Lanchester Parish

Not all birds ever recorded in the parish are listed. There have been rarities such as the Golden Eagle which frequented the western end of the parish for a few days in July 1981 and birds like Raven and Osprey which are occasionally seen overflying the parish. The list attempts to record those birds which might be seen by the informed observer, when they might be seen and the likelihood of seeing them.

Common name	Status in Parish
Little Grebe	Winter visitor
Cormorant	Vagrant
Heron	Common resident
Mute Swan	Vagrant
Greylag Goose	Vagrant
Canada Goose	Vagrant
Pink Footed Goose	Vagrant
Reedbunting	Resident
Mallard	Common resident
Swallow	Common summer resident
House Martin	Common summer resident
Tree Pipit	Summer resident
Meadow Pipit	Common resident
Grey Wagtail	Resident
Pied Wagtail	Common resident
Yellow Wagtail	Summer visitor
Waxwing	Winter visitor
Dipper	Common resident
Sand Martin	Summer resident
Goosander	Resident
Osprey	Summer visitor
Red Kite	Vagrant
Buzzard	Resident
Marsh Harrier	Rare vagrant
Hen Harrier	Rare vagrant
Goshawk	Scarce resident
Sparrowhawk	Common resident
Kestrel	Common resident
Merlin	Vagrant
Hobby	Rare summer visitor
Peregrine Falcon	Vagrant
Red Grouse	Resident
Black Grouse	Vagrant
Pheasant	Common resident
Quail	Summer visitor
Red - legged Partridge	Resident
Grey Partridge	Common resident
Water Rail	Winter visitor
Moorhen	Common resident
Oystercatcher	Summer resident
Golden Plover	Winter visitor
Lapwing	Common resident
Jack Snipe	Winter visitor
Snipe	Common resident
Woodcock	Resident
Curlew	Common resident
Redshank	Summer visitor
Greenshank	Rare vagrant
Green Sandpiper	Rare vagrant
Common Sandpiper	Summer visitor
Black-headed Gull	Common non breeder
Common Gull	Common winter visitor
Lesser Black-backed Gull	Summer visitor
Greater Black-backed Gull	Non breeding resident
Herring Gull	Common vagrant
Collared Dove	Common resident
Stock Dove	Common resident
Woodpigeon	Common resident
Cuckoo	Summer resident
Barn Owl	Resident
Tawny Owl	Common resident
Wren	Common resident
Dunnock	Common resident
Robin	Common resident
Redstart	Summer resident
Winchat	Summer visitor
Stonechat	Scarce resident
Wheatear	Summer resident
Ring Ouzel	Summer vagrant
Blackbird	Common resident
Fieldfare	Common winter visitor
Song Thrush	Common resident
Redwing	Common winter visitor
Mistle Thrush	Common resident
Grasshopper Warbler	Summer resident
Sedge Warbler	Summer resident
Lesser Whitethroat	Summer resident
Whitethroat	Common summer resident
Garden Warbler	Common summer resident
Blackcap	Common summer resident
Wood Warbler	Rare summer visitor
Chiffchaff	Common summer resident
Willow Warbler	Common summer resident
Goldcrest	Common resident
Spotted Flycatcher	Summer resident
Pied Flycatcher	Summer resident
Marsh Tit	Rare resident
Willow Tit	Resident
Coal Tit	Common resident
Blue Tit	Common resident
Great Tit	Common resident
Long-tailed Tit	Common resident
Nuthatch	Resident
Treecreeper	Resident
Jay	Common resident
Magpie	Common resident
Jackdaw	Common resident
Rook	Common resident
Carrion Crow	Common resident
Starling	Common resident
House Sparrow	Resident
Tree Sparrow	Resident
Chaffinch	Common resident
Brambling	Winter visitor

Herons, photo Darin Smith

Long-eared Owl	Scarce resident
Short-eared Owl	Winter visitor
Little Owl	Resident
Swift	Summer resident
Kingfisher	Resident
Green Woodpecker	Resident
Great-spotted Woodpecker	Common resident
Lesser-spotted Woodpecker	Rare vagrant
Skylark	Common resident
Greenfinch	Resident
Goldfinch	Resident
Siskin	Resident
Linnet	Resident
Lesser Redpoll	Resident
Twite	Rare winter visitor
Crossbill	Resident
Bullfinch	Common resident
Yellow Hammer	Common resident

Today birds which Fawcett described as fast disappearing from the countryside such as the Nuthatch (Sitta europaea) have increased in numbers, while casual visitors such as Waxwings (Ampelis garrulous) continue to appear erratically.

Oystercatcher, photo Darin Smith

Dipper, photo Darin Smith

Tawny Owl, photo Darin Smith

25

Sand Martin, photo Darin Smith

Sparrowhawk, photo Darin Smith

Short-eared Owl, photo Darin Smith

Tree Sparrow, photo Darin Smith

Reed Bunting, photo Darin Smith

Meadow Pipit, photo Darin Smith

Yellowhammer, photo Darin Smith

Mammals

by Terry Coult

Mammals are one of the few groups to have a recorded history in Lanchester Parish, principally because some mammal species were perceived to be in direct conflict with human needs and were therefore killed or were hunted for sport. Later, local naturalists published their records in regional journals providing further specialist knowledge of the parish's mammals. As records are rarely parish specific, for the purpose of this text, records in or very close to the parish have been utilised. It is likely that other species occurring, or formerly occurring in the county, will be or have been present in the parish.

Way back in Tudor times it became the responsibility of churchwardens to pay bounty on animals which were perceived to be a threat to human resources. In 1647 the churchwardens of Lanchester decreed that anyone who nailed a fox's head to the church door was to receive of the parish 2 shillings per Fox. Subsequent parish records contain lists of vermin killed, mostly Foxes but in 1652 twelve pence was paid for two Badger heads and again on April the 16th 1661 six pence was paid for a Badger's head. On November the 29th 1760 a Mr William Walton was paid eight pence for a Pine Marten's head. As parish records go the Lanchester Parish records are not all that informative perhaps the churchwardens did not like to pay out or perhaps they just did not keep good records. The adjoining Witton Gilbert Parish records are more detailed and add Otter and Polecat to the list of mammals on which churchwardens paid bounty in the Browney Valley. The Witton Gilbert Parish magazine from September 1901 records that Polecat, Badger and Otter were then still present in the valley.

Around 1847 a hunting pack was formed called the Castleside Hounds, their purpose originally was to hunt the roe deer found in the large woodlands between Salter's Gate and the Derwent Valley. This was at a time when contemporary nineteenth century writers believed that roe were extinct in England, it seems likely that Roe Deer have always had a foothold in north west Durham and that once the Forestry Commission began large scale post war tree planting numbers expanded accordingly and the roe is now quite common in the parish.

In 1840, John Hutchinson of Lanchester (1797-1855), began his manuscript "Durham Fishes, Reptiles and Quadrupeds" which was destined never to be published. The text does however contain some interesting observations on the parish's mammals.

Badger, photo Darin Smith

Brown Hare, photo Darin Smith

Field Vole, photo Darin Smith

Hedgehog, photo Darin Smith

It includes records of Pine Marten from near Butsfield, Waterhouses and Cornsay, Long Eared Bat from Greencroft, Hedgehog from Lanchester and Weasel and Hare from Upper Houses Farm, Lanchester.

J. W. Fawcett the Satley naturalist wrote a series of nature notes in the "Newcastle Weekly Chronicle". On March 1st 1890 his topic was "The Animals of County Durham" and his text includes records of Red Squirrel from Salter's Gate and Black Banks and a cream coloured Mole from Satley. Continuing the theme on the 8th of March 1890 he records Water Shrew from Satley. During the 1914-18 war Fawcett sent a post card to George Bolam the Northumberland naturalist and author of several papers on bats, recording the presence of the Whiskered Bat at Satley and Knitsley.

Not all of these mammals are still with us, the Polecat and Pine Marten are now extinct in the county and the Red Squirrel is extinct in the parish although it still does retain a precarious presence in the county.

In the early 1980s Lanchester Wildlife Group produced a "Provisional List of the Mammals of Lanchester" recording those mammals which could then be found around the village. It included:

Common name	Specific name	Common name	Specific name
Commmon Pipistrelle	*Pipistrellus pipistrellus*	Wood Mouse	*Apodemus sylvaticus*
Whiskered Bat	*Myotis mystacinus*	House Mouse	*Mus musculus*
Noctule Bat	*Nyctalis noctula*	Brown Rat	*Rattus norvegicus*
Long Eared Bat	*Plecotus auritus*	Red Squirrel	*Sciurus vulgaris*
Hedgehog	*Erinaceus europaeus*	Rabbit	*Oryctolagus cuniculus*
Mole	*Talpa europaea*	Hare	*Lepus capensis*
Common Shrew	*Sorex araneus*	Stoat	*Mustela erminea*
Pygmy Shrew	*Sorex minutus*	Weasel	*Mustela nivalis*
Water Shrew	*Neomys fodiens*	Badger	*Meles meles*
Field Vole	*Microtus agrestis*	American Mink	*Mustela vison*
Bank Vole	*Clethrionomys glareolus*	Fox	*Vulpes vulpes*
Water Vole	*Arvicola amphibius*	Roe Deer	*Capreolus capreolus*

Otter, photo Darin Smith

Weasel, photo Darin Smith

Grey Squirrel, photo Darin Smith

Wood Mouse, photo Sue Charlton

Stoat, photo Darin Smith

Bank Vole, photo Darin Smith

29

The list was produced just too early to catch the invasion of the Grey Squirrel but it did pick up the first of the invading American Mink; the Water Vole and Red Squirrel had not yet been lost from the parish. Also in the 1980s there was a small introduction of the Muntjac Deer (*Muntiacus reevesi*) which seems to have failed, although it is possible that this tiny deer has reached the parish as it spreads from the south into the county from more successful releases. Concomitant with the arrival of the Mink was the decline and eventual extinction of the Water Vole within the parish, although like the Red Squirrel it can still be found in the county; both the Water Vole and Red Squirrel are now under threat of national extinction.

Subsequent changes in the parish's mammal fauna add Otter which colonised Durham in the 1990s and now breeds within the parish, as well as Natterer's, Daubenton's and Brandt's Bat to the parish list. In 1999 the Pipistrelle Bat was separated into two species, the Common Pipistrelle and the Soprano Pipistrelle, both of which are found in the parish.

An up to date list of those wild mammals which can confidently be said to breed in the parish should include:

Common name	Specific name	Common name	Specific name
Hedgehog	*Erinaceus europaeus*	Grey Squirrel	*Sciurus carolinensis*
Mole	*Talpa europaea*	Bank Vole	*Clethrionomys glareolus*
Common Shrew	*Sorex araneus*	Field Vole	*Microtus agrestis*
Pygmy Shrew	*Sorex minutus*	Brown Rat	*Rattus norvegicus*
Water Shrew	*Neomys fodiens*	Wood Mouse	*Apodemus sylvaticus*
Daubenton's Bat	*Myotis daubentoni*	House Mouse	*Mus musculus*
Brandt's Bat	*Myotis brandti*	Fox	*Vulpes vulpes*
Noctule	*Nyctalis noctula*	Stoat	*Mustela erminea*
Commmon Pipistrelle	*Pipistrellus pipistrellus*	Weasel	*Mustela nivalis*
Soprano Pipistrelle	*Pipistrellus pygmaeus*	American Mink	*Mustela vison*
Long Eared Bat	*Plecotus auritus*	Badger	*Meles meles*
Whiskered Bat	*Myotis mystacinus*	Otter	*Lutra lutra*
Rabbit	*Oryctolagus cuniculus*	Roe Deer	*Capreolus capreolus*
Brown Hare	*Lepus capensis*		

Roe Deer, photo Darin Smith

Rabbit, photo Darin Smith

American Mink, photo Darin Smith

Fish

by Terry Coult

Quantifiable fish data is sparse for the River Browney and its tributaries with only the Environment Agency carrying out regular sampling. In the Lanchester Parish sampling takes place at the Hythe, Partridge Close and near Square House. The Smallhope Burn is sampled in Lanchester itself. Results for wild fish within the parish gathered between 2003 and 2010 include Atlantic Salmon, Brown Trout, Sea Trout, European Eel, Bullhead, Stoneloach, Minnow and Brook Lamprey. Casual records for Malton include Minnow, Stoneloach, Bullhead, Brown Trout and Brook Lamprey which has also been recorded on the Knitsley Burn above Knitsley Mill. Analysis of otter spraint from the Browney in 1998 showed that, in order of preference, otters were eating salmonids (Trout and Salmon), Bullhead, Eel, Stoneloach and Minnow.

There are a number of stocked stillwaters along the River Browney which hold Rainbow Trout (*Oncorhyncus mykiss*), Perch (*Perca sp.*), Tench (*Tinca tinca*), Gudgeon (*Gobio gobio*), Dace (*Leuciscus leuciscus*), Common Bream (*Abramis brama*) and Carp (*Cyprinus sp.*). Within the parish, Knitsley Mill Fishery stocks Rainbow Trout and Lizards Fishery near Lanchester has Brown Trout and unspecified coarse fish. Fishing clubs along the Browney supplement the wild stock with additional Brown Trout and Grayling (*Thymallus thymallus*) and the lower end of the river will have natural populations of Chub (*Squalius aphalus*) and Barbel (*Barbus barbus*).

Salmon, photo Joe Ridley

Environment Agency Fish Data

River Browney, Hythe, NZ1160044800
2009

Common name	Specific name
Brown Trout	*Salmo trutta*
Sea Trout	*Salmo trutta*
Bullhead	*Cottus gobio*
Stoneloach	*Noemacheilus barbatulus*

2008
Brown Trout	*Salmo trutta*
Sea Trout	*Salmo trutta*
European Eel	*Anguilla anguilla*
Stoneloach	*Noemacheilus barbatulus*
Bullhead	*Cottus gobio*

2007
Brown Trout	*Salmo trutta*
Sea Trout	*Salmo trutta*
Bullhead	*Cottus gobio*
Stoneloach	*Noemacheilus barbatulus*

Perch

Bullhead

River Browney, Partridge Close, NZ1430044900

2010
Brown Trout   *Salmo trutta*
Sea Trout   *Salmo trutta*
Bullhead   *Cottus gobio*
Brook Lamprey   *Lampetra planeri*

2009
Brown Trout   *Salmo trutta*
Sea Trout   *Salmo trutta*

2008
Brown Trout   *Salmo trutta*
Sea Trout   *Salmo trutta*
Bullhead   *Cottus gobio*
Stoneloach   *Noemacheilus barbatulus*

2007
Atlantic Salmon   *Salmo salar*
Brown Trout   *Salmo trutta*
Sea Trout   *Salmo trutta*
Bullhead   *Cottus gobio*
Stoneloach   *Noemacheilus barbatulus*

River Browney, Square House, NZ1580045300

Brown Trout   *Salmo trutta*
Sea Trout   *Salmo trutta*

Smallhope Burn, Lanchester, NZ

2003
Brown Trout   *Salmo trutta*
Minnow   *Phoxinus phoxinus*
Stoneloach   *Noemacheilus barbatulus*

Brown Trout

Chub

Grayling

Reptiles and Amphibians

by Rachel Jackson and Terry Coult

Reptiles
There are few written records for reptiles within the parish. Writing in "The Naturalist" No. 517, in February 1901, J. W. Fawcett records Grass Snake (*Natrix natrix*) under its old name of Ringed Snake at West Butsfield in 1883 and at Satley (just outside the parish) in 1886. In the same publication in July 1901, No. 534, he records the Ringed Snake at East Butsfield in 1900. There are no further records for this snake in the parish and it is now extremely rare, possibly verging on extinction in the county. The western end of the parish still supports Common Lizard (*Zootoca vivipara*), Adder (*Vipera berus*) and Slow Worm (*Anguis fragilis*) although it is likely that as habitats are degraded through neglect, afforestation and agricultural improvement these species are declining.

Amphibians
The Ordnance Survey shows 35 ponds fairly evenly scattered across the parish with several more close to the parish boundary. It is likely that some of these ponds will have been lost through time in line with the national trend of loss of countryside ponds through agricultural improvement. A few new ponds have been created within the parish and there will be garden ponds which are not mapped, so it is possible that the parish pond resource remains quite healthy. Some ponds will have been stocked with fish, which limits their value for wildlife and will eventually lead to the loss of the amphibian population, with the exception of Common Toad (*Bufo bufo*). Few of these ponds have been surveyed, the most regularly recorded will be Durham Wildlife Trust's Malton Nature Reserve ponds which supports five native British amphibians, Common Frog (*Rana temporaria*), Common Toad, Smooth Newt (*Lissotriton vulgaris*), Palmate Newt (*Lissotriton helveticus*) and Great Crested Newt (*Triturus cristatus*). The cluster of ponds in the

Toad, photo Sue Charlton

Adder, photo Sue Charlton

Great Crested Newt, photo Stuart Priestley

Common Lizard, photo Darin Smith

disused quarry at Quickburn, which is just outside the western end of the parish, also supports the same suite of amphibians, so they can be said to be found at both ends of the parish. What is not known is their status in the ponds between. There is a single record of a Great Crested Newt at Woodlands Hall but until systematic survey work is carried out the status of the parish's amphibian population remains mostly unknown.

Whilst maps show a good spread of ponds across the parish they do not include the many garden ponds which provide an important distribution network for amphibians. Due to the loss of countryside ponds, garden ponds assume a much greater value as amphibian habitat.

All of the parish's reptiles and amphibians are under threat and in decline as a result of agricultural and farming practice changes, the draining of ponds, stocking ponds with fish and the general trend to "tidy" the countryside.

Slow Worm, photo Terry Coult

Invertebrates

by Terry Coult

There are few historical invertebrate records for the Lanchester Parish, with the exception of butterflies and moths which were once the passion of collectors and therefore reasonably well recorded historically and in the current day. Otherwise, both historical and current records for many invertebrate groups are scarce and so casual as to make them valueless as indicators or measures of change within the parish. This section therefore deals with invertebrate groups which are well known to local naturalists and the public and/or have sufficient records to justify their inclusion. Some groups like the moths have so many records that they can't be accommodated in a document such as this; there are over two and a half thousand moth records for the parish alone and as a result this text does not seek to record every single invertebrate but hopefully records all species within each group which have been recorded in the parish. Durham has a regional records centre, the Environmental Records Information Centre (ERIC), based at the Great North Museum Hancock in Newcastle upon Tyne and anyone wanting details of records can find them there.

The Lanchester Parish has a diverse range of habitats ranging from upland moorlands and heathland in the west, through farmland, deciduous and conifer woodland, wetland, scrub, hedgerow and species rich grasslands As a result it has a very diverse invertebrate fauna including species peculiar to all habitats.

Dragonflies

Like moths and butterflies there has been an increase in dragonfly species recorded in the parish and county over the last several years, mostly southern species moving north. There has also been an increase in migrant species reaching the parish. Dragonflies like the Broad-bodied Chaser and the Four-spotted Chaser are recent arrivals to breed and there has been an increase in migrant species like the Migrant Hawker reaching the parish.

The dragonfly family is separated into the damselflies (*Zygoptera*) which are generally smaller with a weak fluttering flight and the dragonflies (*Anisoptera*) much bigger flies, with a strong flight. Those species recorded in the Lanchester Parish are:

Common name	Specific name	Status in Parish
Banded Demoiselle	*Calopteryx splendens*	Rare vagrant
Emerald Damselfly	*Lestes sponsa*	Breeding
Large Red Damselfly	*Pyrrhosoma nymphula*	Breeding
Azure Damselfly	*Coenagrion puella*	Breeding
Common Blue Damselfly	*Enallagma cyathigerum*	Breeding
Blue-tailed Damselfly	*Ischnura elegans*	Breeding
Common Hawker Dragonfly	*Aeshna juncea*	Breeding
Migrant Hawker Dragonfly	*Aeshna mixta*	Migrant visitor
Southern Hawker Dragonfly	*Aeshna cyanea*	Breeding
Emperor Dragonfly	*Anax imperator*	Vagrant
Golden-ringed Dragonfly	*Cordulegaster boltonii*	Breeding
Four-spotted Chaser Dragonfly	*Libellula quadrimaculata*	Breeding
Broad-bodied Chaser Dragonfly	*Libellula depressa*	Breeding
Common Darter Dragonfly	*Sympetrum striolatum*	Breeding
Ruddy Darter Dragonfly	*Sympetrum sanguineum*	Breeding
Black Darter Dragonfly	*Sympetrum danae*	Probable breeder
Yellow-winged Darter Dragonfly	*Sympetrum flaveolum*	Rare migrant visitor
Red-veined Darter Dragonfly	*Sympetrum fonscolombii*	Rare migrant visitor

Migrant Hawker Dragonfly, photo Darin Smith

Large Red Damselfy, photo Sue Charlton

Ruddy Darter Dragonfly, photo Darin Smith

Broad Bodied Chaser Dragonfly, photo Dean Heward

Golden Ring Dragonfly in wheel, photo Terry Coult

Hoverflies
Hoverfly records are mainly confined to the extreme east of the parish and having no common names are not easy for the general public to relate to but as so many have been recorded in the parish the list is included.

Baccha sp
Melanostoma mellinum
Melanostoma scalare
Platycheirus albimanus
Platycheirus angustatus
Platycheirus clypeatus
Platycheirus manicatus
Platycheirus scambus
Platycheirus scutatus
Pryophaena granditarsa
Pryophaena rosarum
Paragus haemorrhous
Chrysotoxum arcuatum
Chrysotoxum bicinctum
Dasysyrphus albostriatus
Dasysyrphus lunulatus
Dasysyrphus tricinctus
Dasysyrphus venustus
Dasysyrphus friulensis
Didea fasciata
Epistrophe eligans
Epistrophe grossulariae
Episyrphus balteatus
Leucozona glaucia
Leucozona lucorum
Megasyrphus annulipes
Melangyna compositarum
Melangyna lasiophthalma
Melangyna quadrimaculata
Melangyna meligramma
Meliscaeva cinctella
Metasyrphus corollae
Metasyrphus latifasciatus
Metasyrphus luniger
Parasyrphus punctulatus
Scaeva pyrastri
Sphaerophora menthastri
Sphaerophora sp.
Syrphus ribesii
Syrphus torvus
Syrphus vitripennis
Cheilosia albitarsis
Cheilosia bergenstammi
Cheilosia grossa
Cheilosia illustrata
Cheilosia pagana
Cheilosia variabilis
Cheilosia vernalis
Cheilosia nebulosa
Ferdinandia cuprea
Portevinia maculata
Rhingia campestris
Chrysogaster hirtella
Chrysogaster solstitialis
Chrysogaster chalybeata
Lejogaster metallina
Spehgina clunipes
Neoascia podagrica
Anasymia contracta
Eristalinus sepulchralis
Eristalis arbustorum
Eristalis horticola
Eristalis intricarius
Eristalis pertinax
Eristalis tenax
Helophilus pendulus
Helophilus trivittatus
Myathropa florea
Merodon equestris
Pipizella varipes
Arctophila fulva
Sericomyia lappona
Sericomyia silentis
Vollucella bombylans
Vollucella pellucens
Criorhina ranunculi
Criorhina berberina
Syritta pipiens
Xylota segnis
Xylota sylvarum

Bumblebees

Most of the bumblebee species recorded in the parish are common and widespread with the exception of *Bombus monticola*, sometimes called the Bilberry Bumblebee because of its close association with the plant. This bumblebee has been found in recent years in the west of the parish and is possibly increasing in numbers. The other species are the common or garden ones, to some extent actually depending on flower rich gardens to support their numbers. Some bumblebees, the Cuckoo Bumblebees, parasitise the nests of other bumblebees killing the queen and replacing her eggs with their own, two of these species have been recorded in the parish. Bumblebee records for the parish are:

Common name	Specific name
Red Tailed Bumblebee	*Bombus lapidarius*
Bilberry Bumblebee	*Bombus monticola*
Early Bumblebee	*Bombus pratorum*
Bufftailed Bumblebee	*Bombus terrestris*
White Tailed Bumblebee	*Bombus lucorum*
Garden Bumblebee	*Bombus hortorum*
Cuckoo Bumblebee	*Bombus bohemicus*
Cuckoo Bumblebee	*Bombus sylvestris*
Common Carder Bumblebee	*Bombus pascuorum*

Shieldbugs

Only a handful of shieldbugs have been recorded in the parish, the rare one being *Picromerus bidens*. The list for the parish is:

Birch Shieldbug	*Elasmostethus interstinctus*
Parent Bug	*Elasmucha grisea*
Hawthorn Shieldbug	*Acanthosoma haemorrhoidale*
Gorse Shieldbug	*Piezodorus lituratus*
Blue Shieldbug	*Zircrona caerulea*
Forest Shieldbug	*Pentatoma rufipes*
Spined Shieldbug	*Picromerus bidens*

Grasshoppers

The list of grasshoppers for the parish is very short:

Common Field Grasshopper	*Chorthippus brunneus*
Mottled Grasshopper	*Myrmeleotettix maculatus*
Meadow Grasshopper	*Chorthippus parallelus*
Common Green Grasshopper	*Omocestus viridulus*

Bilberry Bumblebee, photo Terry Coult

Spined Shieldbug, photo Stuart Priestley

Buff Tailed Bumblebee, photo Darin Smith

Common Field Grasshopper, photo Darin Smith

Common Green Grasshopper, photo Darin Smith

Butterflies

J. W. Fawcett published "A History of the Parish of Dipton" in 1911. Within it is a list of "The Butterflies of Dipton and District" supplied by Mr Thomas Gatiss. His list extends into the Lanchester Parish and provides early records of Orange Tip, Common Blue, Dark Green Fritillary, Pearl Bordered Fritillary and Dingy Skipper. Writing in "The Vasculum" Vol. XX, No. 3 in August 1934, J. W. Heslop Harrison under the heading "Three Notable Days" records a July 1st visit to Lanchester where he and his colleagues found Small Heath, Meadow Brown, Small Pearl Bordered Fritillary and Pearl Bordered Fritillary. The latter two butterflies described as in their thousands at just this one site.

Today the Lanchester Parish holds almost all of the last few of the Small Pearl Bordered Fritillary colonies in the county. The Pearl Bordered Fritillary is extinct in the county and is in severe decline nationally. The Dark Green Fritillary remains in the parish and the county but in much reduced numbers. All of the other species recorded by Gatiss and Heslop Harrison are declining with the Dingy Skipper possibly declining the most quickly of all.

In recent years the Purple Hairstreak and the Speckled Wood have been newly recorded in the parish. The former may just have been overlooked as it can be hard to find but the latter is a butterfly returning to the north of England and to Lanchester Parish after over a hundred years of absence.

As a result of the diversity of habitats and because the parish has not suffered too severely from agricultural intensification the list of butterfly species breeding in the parish is still good, containing:

Common name	Specific name	Common name	Specific name
Small Skipper	Thymelicus sylvestris	Red Admiral	Vanessa atalanta
Large Skipper	Ochlodes venata	Painted Lady	Cynthia cardui
Dingy Skipper	Erynnis tages	Small Tortoiseshell	Aglais urticae
Large White	Pieris brassicae	Peacock	Inachis io
Small White	Pieris rapae	Comma	Polygonia c-album
Green Veined White	Pieris napi	Small Pearl Bordered Fritillary	Boloria selene
Orange Tip	Anthocharis cardamines	Dark Green Fritillary	Argynnis aglaja
Green Hairstreak	Callophrys rubi	Speckled Wood	Pararge aegeria
Purple Hairstreak	Quercusia quercus	Wall	Lasiommata megera
White Letter Hairstreak	Strymonidia w-album	Meadow Brown	Maniola jurtina
Small Copper	Lycaena phlaeas	Small Heath	Coenonympha pamphilus
Common Blue	Polyommatus icarus	Ringlet	Aphantopus hyperantus
Holly Blue	Celastrina argiolus		

As well as the breeding species there are a few none residents which occasionally visit the parish. Brimstone (*Gonepteryx rhamni*) occasionally wanders north from Yorkshire into the county and has been seen in the parish and both Clouded Yellow (*Colias croceus*) and the Camberwell Beauty (*Nymphalis antiopa*) have been recorded in the parish as migrants, the latter very rarely.

Common Blue butterfly, photo Darin Smith

Large Skipper butterfly, photo Darin Smith

Orange Tip butterfly, photo Darin Smith

Green Hairstreak Butterfly, photo Darin Smith

Small Pearl Bordered Fritillary, photo Sue Charlton

Wall butterfly, photo Darin Smith

Comma butterfly, photo Darin Smith

Small Skipper butterfly, photo Darin Smith

Lunar Hornet Clearwing, photo Stuart Priestley

Moths

Like the butterflies the parish has a diversity of moth species because it incorporates so many habitats. It has however only two real claims to fame in the moth world, the Large Red Belted Clearwing moth and the Lead Coloured Drab both of which are very rare. Over the last few years moths such as the Red Underwing, Svensson's Copper Underwing and Blair's Shoulder Knot have colonised the county and parish from the south. Whether this is a reflection of global warming is unknown but it is true that some resident moth species are changing their flight times and appearing earlier or later in the year.

Moths are divided into macro and micro moths and the micro moths do not generally have common names. The following list includes all 574 moth species ever recorded in the parish. What it doesn't do is record location or status, ERIC is the place to find out those details.

Common Name	Specific name	Common Name	Specific name
	Micropterix calthella	Large Emerald	*Geometra papilionaria*
	Eriocrania subpurpurella	Small Fan-footed Wave	*Idaea biselata*
	Eriocrania unimaculella	Small Dusty Wave	*Idaea seriata*
	Eriocrania sparrmannella	Single-dotted Wave	*Idaea dimidiata*
	Eriocrania sangii	Riband Wave	*Idaea aversata*
	Eriocrania semipurpurella	Flame Carpet	*Xanthorhoe designata*
Ghost Moth	*Hepialus humuli*	Red Carpet	*Xanthorhoe decoloraria*
Orange Swift	*Hepialus sylvina*	Silver-ground Carpet	*Xanthorhoe montanata*
Gold Swift	*Hepialus hecta*	Garden Carpet	*Xanthorhoe fluctuata*
Common Swift	*Hepialus lupulinus*	Shaded Broad-bar	*Scotopteryx chenopodiata*
Map-winged Swift	*Hepialus fusconebulosa*	July Belle	*Scotopteryx luridata*
	Ectoedemia atricollis	Small Argent & Sable	*Epirrhoe tristata*
	Ectoedemia occultella	Common Carpet	*Epirrhoe alternata*
	Ectoedemia minimella	Yellow Shell	*Camptogramma bilineata*
	Ectoedemia albifasciella	Grey Mountain Carpet	*Entephria caesiata*
	Trifurcula immundella	Shoulder Stripe	*Anticlea badiata*
	Stigmella aurella	Streamer	*Anticlea derivata*
	Stigmella sorbi	Beautiful Carpet	*Mesoleuca albicillata*
	Stigmella plagicolella	Dark Spinach	*Pelurga comitata*
	Stigmella salicis	Water Carpet	*Lampropteryx suffumata*
	Stigmella obliquella	Purple Bar	*Cosmorhoe ocellata*
	Stigmella trimaculella	Chevron	*Eulithis testata*
	Stigmella floslactella	Northern Spinach	*Eulithis populata*
	Stigmella tityrella	Spinach	*Eulithis mellinata*
		Barred Straw	*Eulithis pyraliata*
	Stigmella perpygmaeella	Small Phoenix	*Ecliptopera silaceata*
	Stigmella hemargyrella		

Common Name	Scientific Name
	Stigmella atricapitella
	Stigmella ruficapitella
Dark Marbled Carpet	*Chloroclysta citrata*
	Stigmella svenssoni
Rose Leaf Miner	*Stigmella anomalella*
	Stigmella hybnerella
	Stigmella oxyacanthella
	Stigmella nylandriella
	Stigmella magdalenae
	Stigmella regiella
	Stigmella crataegella
	Stigmella betulicola
	Stigmella microtheriella
	Stigmella alnetella
	Stigmella lapponica
	Stigmella confusella
	Tischeria ekebladella
	Emmetia marginea
	Incurvaria praelatella
	Nematopogon swammerdamella
	Nematopogon schwarziellus
	Nemophora degeerella
	Adela reaumurella
Six-spot Burnet	*Zygaena filipendulae*
Narrow-bordered Five-spot Burnet	*Zygaena lonicerae latomarginata*
Cork Moth	*Nemapogon cloacella*
	Triaxomera fulvimitrella
	Monopis weaverella
	Monopis fenestratella
Common Clothes Moth	*Tineola bisselliella*
	Tinea flavescentella
	Tinea semifulvella
	Tinea trinotella
	Ochsenheimeria urella
Apple Leaf Miner	*Lyonetia clerkella*
	Caloptilia elongella
	Caloptilia betulicola
	Caloptilia rufipennella
	Caloptilia alchimiella
	Caloptilia syringella
	Aspilapteryx tringipennella
	Eucalybites auroguttella
Red-green Carpet	*Chloroclysta siterata*
Autumn Green Carpet	*Chloroclysta miata*
Common Marbled Carpet	*Chloroclysta truncata*
Barred Yellow	*Cidaria fulvata*
Pine Carpet	*Thera firmata*
Grey Pine Carpet	*Thera obeliscata*
Spruce Carpet	*Thera britannica*
Juniper Carpet	*Thera juniperata*
Broken-barred Carpet	*Electrophaes corylata*
Beech-green Carpet	*Colostygia olivata*
Mottled Grey	*Colostygia multistrigaria*
Green Carpet	*Colostygia pectinataria*
July Highflyer	*Hydriomena furcata*
May Highflyer	*Hydriomena impluviata*
November Moth agg.	*Epirrita dilutata agg.*
Autumnal Moth	*Epirrita autumnata*
Small Autumnal Moth	*Epirrita filigrammaria*
Winter Moth	*Operophtera brumata*
Northern Winter Moth	*Operophtera fagata*
Rivulet	*Perizoma affinitata*
Small Rivulet	*Perizoma alchemillata*
Grass Rivulet	*Perizoma albulata*
Sandy Carpet	*Perizoma flavofasciata*
Twin-spot Carpet	*Perizoma didymata*
Slender Pug	*Eupithecia tenuiata*
Toadflax Pug	*Eupithecia linariata*
Foxglove Pug	*Eupithecia pulchellata*
Marbled Pug	*Eupithecia irriguata*
Marsh Pug	*Eupithecia pygmaeata*
Triple-spotted Pug	*Eupithecia trisignaria*
Wormwood Pug	*Eupithecia absinthiata*
Common Pug	*Eupithecia vulgata*
White-spotted Pug	*Eupithecia tripunctaria*
Grey Pug	*Eupithecia subfuscata*
Narrow-winged Pug	*Eupithecia nanata*
Brindled Pug	*Eupithecia abbreviata*
Dwarf Pug	*Eupithecia tantillaria*
Green Pug	*Pasiphila rectangulata*
Streak	*Chesias legatella*
Manchester Treble-bar	*Carsia sororiata*
Chimney Sweeper	*Odezia atrata*

Silver-Y Moth, photo Darin Smith

Common Heath Moth, photo Darin Smith

Chimney Sweeper Moth, photo Darin Smith

	Parornix anglicella	Welsh Wave	*Venusia cambrica*	
	Parornix devoniella	Small Yellow Wave	*Hydrelia flammeolaria*	
	Deltaornix torquillella	Early Tooth-striped	*Trichopteryx carpinata*	
	Phyllonorycter roboris	Yellow-barred Brindle	*Acasis viretata*	
	Phyllonorycter quercifoliella			
	Phyllonorycter messaniella	Magpie Moth	*Abraxas grossulariata*	
	Phyllonorycter oxyacanthae	Clouded Border	*Lomaspilis marginata*	
	Phyllonorycter sorbi	Tawny-barred Angle	*Macaria liturata*	
	Phyllonorycter blancardella	Latticed Heath	*Chiasmia clathrata*	
	Phyllonorycter maestingella	V-Moth	*Macaria wauaria*	
	Phyllonorycter rajella	Brown Silver-line	*Petrophora chlorosata*	
	Phyllonorycter nigrescentella	Barred Umber	*Plagodis pulveraria*	
	Phyllonorycter ulmifoliella	Brimstone Moth	*Opisthograptis luteolata*	
	Phyllonorycter nicellii	Lilac Beauty	*Apeira syringaria*	
Lunar Hornet Moth	*Sesia bembeciformis*	Canary-shouldered Thorn	*Ennomos alniaria*	
Large Red-belted Clearwing	*Synanthedon culiciformis*	Dusky Thorn	*Ennomos fuscantaria*	Large Red Belted Clearwing, photo Keith Dover
	Anthophila fabriciana	Early Thorn	*Selenia dentaria*	
Cocksfoot Moth	*Glyphipterix simpliciella*	Lunar Thorn	*Selenia lunularia*	
	Glyphipterix fuscoviridella	Scalloped Hazel	*Odontopera bidentata*	
	Glyphipterix thrasonella	Scalloped Oak	*Crocallis elinguaria*	
	Argyresthia brockeella	Swallow-tailed Moth	*Ourapteryx sambucaria*	
	Argyresthia goedartella	Feathered Thorn	*Colotois pennaria*	
	Argyresthia sorbiella	Pale Brindled Beauty	*Phigalia pilosaria*	
	Argyresthia curvella	Oak Beauty	*Biston strataria*	
Apple Fruit Moth	*Argyresthia conjugella*	Peppered Moth	*Biston betularia*	
	Argyresthia semifusca	Spring Usher	*Agriopis leucophaearia*	Oak Beauty Moth, photo Terry Coult
Bird-cherry Ermine	*Yponomeuta evonymella*	Scarce Umber	*Agriopis aurantiaria*	
Orchard Ermine	*Yponomeuta padella*	Dotted Border	*Agriopis marginaria*	
	Swammerdamia compunctella	Mottled Umber	*Erannis defoliaria*	
	Paraswammerdamia nebulella	Willow Beauty	*Peribatodes rhomboidaria*	
	Cedestis gysseleniella	Mottled Beauty	*Alcis repandata*	
	Cedestis subfasciella	Engrailed	*Ectropis bistortata*	
Honeysuckle Moth	*Ypsolopha dentella*	Grey Birch	*Aethalura punctulata*	
	Ypsolopha scabrella	Common Heath	*Ematurga atomaria*	
	Ypsolopha parenthesella	Bordered White	*Bupalus piniaria*	
	Ypsolopha ustella	Common White Wave	*Cabera pusaria*	
Diamond-back Moth	*Plutella xylostella*	Common Wave	*Cabera exanthemata*	
	Plutella porrectella	Clouded Silver	*Lomographa temerata*	
	Epermenia chaerophyllella	Early Moth	*Theria primaria*	
	Schreckensteinia festaliella	Light Emerald	*Campaea margaritata*	
	Coleophora lutipennella	Barred Red	*Hylaea fasciaria*	
	Coleophora gryphipennella	Grey Scalloped Bar	*Dyscia fagaria*	
	Coleophora flavipennella	Poplar Hawk-moth	*Laothoe populi*	5 Spot Burnet Moth, photo Darin Smith

Larch Case-bearer	*Coleophora serratella*	Humming-bird Hawk-moth	*Macroglossum stellatarum*
	Coleophora mayrella	Bedstraw Hawk-moth	*Hyles gallii*
	Coleophora laricella	Elephant Hawk-moth	*Deilephila elpenor*
	Coleophora lixella	Small Elephant Hawk-moth	*Deilephila porcellus*
	Coleophora albicosta	Buff-tip	*Phalera bucephala*
	Coleophora discordella	Puss Moth	*Cerura vinula*
	Coleophora caespititiella	Sallow Kitten	*Furcula furcula*
	Elachista atricomella	Poplar Kitten	*Furcula bifida*
	Elachista luticomella	Iron Prominent	*Notodonta dromedarius*
	Elachista albifrontella	Pebble Prominent	*Notodonta ziczac*
	Elachista canapennella	Lesser Swallow Prominent	*Pheosia gnoma*
	Elachista rufocinerea	Swallow Prominent	*Pheosia tremula*
	Elachista cerusella	Coxcomb Prominent	*Ptilodon capucina*
	Elachista argentella	Scarce Prominent	*Odontosia carmelita*
Brown House Moth	*Hofmannophila pseudospretella*	Pale Prominent	*Pterostoma palpina*
White-shouldered House Moth	*Endrosis sarcitrella*	Lunar Marbled Brown	*Drymonia ruficornis*
	Carcina quercana	Figure of Eight	*Diloba caeruleocephala*
	Diurnea fagella	Vapourer	*Orgyia antiqua*
	Agonopterix heracliana	Muslin Footman	*Nudaria mundana*
	Agonopterix ciliella	Red-necked Footman	*Atolmis rubricollis*
	Agonopterix alstromeriana	Common Footman	*Eilema lurideola*
	Agonopterix arenella	Wood Tiger	*Parasemia plantaginis*
	Agonopterix ocellana	Garden Tiger	*Arctia caja*
	Agonopterix assimilella	White Ermine	*Spilosoma lubricipeda*
	Agonopterix angelicella	Buff Ermine	*Spilosoma luteum*
	Metzneria metzneriella	Muslin Moth	*Diaphora mendica*
	Eulamprotes atrella	Ruby Tiger	*Phragmatobia fuliginosa*
	Exoteleia dodecella	Cinnabar	*Tyria jacobaeae*
	Carpatolechia notatella	Short-cloaked Moth	*Nola cucullatella*
	Carpatolechia proximella	Least Black Arches	*Nola confusalis*
	Teleiopsis diffinis	White-line Dart	*Euxoa tritici*
	Bryotropha affinis	Garden Dart	*Euxoa nigricans*
	Bryotropha terrella	Turnip Moth	*Agrotis segetum*
	Mirificarma mulinella	Heart and Club	*Agrotis clavis*
	Aroga velocella	Heart and Dart	*Agrotis exclamationis*
	Neofaculta ericetella	Dark Sword-grass	*Agrotis ipsilon*
	Caryocolum fraternella	Shuttle-shaped Dart	*Agrotis puta*
	Caryocolum blandella	Flame Shoulder	*Ochropleura plecta*
	Caryocolum tricolorella	Dotted Rustic	*Rhyacia simulans*
	Syncopacma sangiella	Large Yellow Underwing	*Noctua pronuba*
	Syncopacma cinctella	Broad-bordered Yellow Underwing	*Noctua fimbriata*
	Acompsia cinerella	Lesser Broad-bordered Yellow Underwing	*Noctua janthe*
	Hypatima rhomboidella	Least Yellow Underwing	*Noctua interjecta*

White Ermine Moth, photo Terry Coult

6 Spot Burnet Moth, photo Terry Coult

Poplar Hawk Moth, photo Terry Coult

	Blastobasis lacticolella	Double Dart	*Graphiphora augur*
	Mompha raschkiella	Autumnal Rustic	*Eugnorisma glareosa*
	Blastodacna hellerella	True Lover's Knot	*Lycophotia porphyrea*
		Ingrailed Clay	*Diarsia mendica*
	Cochylimorpha straminea	Barred Chestnut	*Diarsia dahlii*
	Agapeta hamana	Purple Clay	*Diarsia brunnea*
	Aethes cnicana	Small Square-spot	*Diarsia rubi*
	Aethes rubigana	Setaceous Hebrew Character	*Xestia c-nigrum*
	Eupoecilia angustana	Triple-spotted Clay	*Xestia ditrapezium*
	Falseuncaria ruficiliana	Double Square-spot	*Xestia triangulum*
Barred Fruit-tree Tortrix	*Pandemis cerasana*		
	Pandemis cinnamomeana	Dotted Clay	*Xestia baja*
Dark Fruit-tree Tortrix	*Pandemis heparana*	Six-striped Rustic	*Xestia sexstrigata*
	Syndemis musculana	Square-spot Rustic	*Xestia xanthographa*
Timothy Tortrix	*Aphelia paleana*	Gothic	*Naenia typica*
	Aphelia unitana	Green Arches	*Anaplectoides prasina*
	Clepsis consimilana		
	Lozotaenia forsterana	Red Chestnut	*Cerastis rubricosa*
	Capua vulgana	Beautiful Yellow Underwing	*Anarta myrtilli*
	Pseudargyrotoza conwagana	Shears	*Hada nana*
	Olindia schumacherana	Cabbage Moth	*Mamestra brassicae*
	Isotrias rectifasciana	Pale-shouldered Brocade	*Lacanobia thalassina*
	Eulia ministrana	Bright-line Brown-eye	*Lacanobia oleracea*
Grey Tortrix	*Cnephasia stephensiana*	Glaucous Shears	*Papestra biren*
Flax Tortrix	*Cnephasia asseclana*	Broom Moth	*Melanchra pisi*
Light Grey Tortrix	*Cnephasia incertana*	Broad-barred White	*Hecatera bicolorata*
	Tortricodes alternella	Campion	*Hadena rivularis*
	Exapate congelatella	Lychnis	*Hadena bicruris*
	Neosphaleroptera nubilana	Antler Moth	*Cerapteryx graminis*
Green Oak Tortrix	*Tortrix viridana*	Feathered Gothic	*Tholera decimalis*
	Acleris forsskaleana	Pine Beauty	*Panolis flammea*
	Acleris laterana	Small Quaker	*Orthosia cruda*
Strawberry Tortrix	*Acleris comariana*	Lead-coloured Drab	*Orthosia populeti*
	Acleris sparsana	Powdered Quaker	*Orthosia gracilis*
Rhomboid Tortrix	*Acleris rhombana*	Common Quaker	*Orthosia cerasi*
Garden Rose Tortrix	*Acleris variegana*	Clouded Drab	*Orthosia incerta*
	Acleris hastiana	Twin-spotted Quaker	*Orthosia munda*
	Acleris literana	Hebrew Character	*Orthosia gothica*
	Acleris emargana	Brown-line Bright Eye	*Mythimna conigera*
	Olethreutes schulziana	Clay	*Mythimna ferrago*
	Olethreutes palustrana	Smoky Wainscot	*Mythimna impura*
Common Wainscot	*Mythimna pallens*		
	Celypha lacunana	Shoulder-striped Wainscot	*Mythimna comma*

Plume Moth, photo Terry Coult

Orange Underwing Moth, photo Terry Coult

Mother Shipton Moth, photo Terry Coult

Plum Tortrix	*Hedya pruniana*	Shark	*Cucullia umbratica*	
Marbled Orchard Tortrix	*Hedya nubiferana*	Minor Shoulder-knot	*Brachylomia viminalis*	
	Orthotaenia undulana	Deep-brown Dart	*Aporophyla lutulenta*	
	Apotomis turbidana	Northern Deep-brown Dart	*Aporophyla lueneburgensis*	
	Apotomis betuletana	Black Rustic	*Aporophyla nigra*	
	Endothenia nigricostana	Pale Pinion	*Lithophane hepatica*	
	Lobesia littoralis	Blair's Shoulder-knot	*Lithophane leautieri*	
	Bactra lancealana	Early Grey	*Xylocampa areola*	
	Ancylis geminana	Green-brindled Crescent	*Allophyes oxyacanthae*	
	Ancylis laetana	Merveille du Jour	*Dichonia aprilina*	
	Ancylis badiana	Brindled Green	*Dryobotodes eremita*	
	Epinotia bilunana	Dark Brocade	*Blepharita adusta*	
	Epinotia ramella	Grey Chi	*Antitype chi*	
	Epinotia immundana	Satellite	*Eupsilia transversa*	
	Epinotia tetraquetrana	Chestnut	*Conistra vaccinii*	
	Epinotia nisella	Dark Chestnut	*Conistra ligula*	
	Epinotia tedella	Brick	*Agrochola circellaris*	
Willow Tortrix	*Epinotia cruciana*	Red-line Quaker	*Agrochola lota*	
	Epinotia trigonella	Yellow-line Quaker	*Agrochola macilenta*	
	Epinotia caprana	Flounced Chestnut	*Agrochola helvola*	
	Epinotia brunnichana	Brown-spot Pinion	*Agrochola litura*	
Spruce Bud Moth	*Zeiraphera ratzeburgiana*	Beaded Chestnut	*Agrochola lychnidis*	
	Zeiraphera isertana	Suspected	*Parastichtis suspecta*	
	Gypsonoma dealbana	Centre-barred Sallow	*Atethmia centrago*	
	Epiblema cynosbatella	Lunar Underwing	*Omphaloscelis lunosa*	
Bramble Shoot Moth	*Epiblema uddmanniana*	Pink-barred Sallow	*Xanthia togata*	
	Epiblema trimaculana	Sallow	*Xanthia icteritia*	
	Epiblema roborana	Poplar Grey	*Acronicta megacephala*	
	Epiblema scutulana	Miller	*Acronicta leporina*	
	Epiblema cirsiana	Alder Moth	*Acronicta alni*	
	Epiblema costipunctana	Dark Dagger / Grey Dagger	*Acronicta tridens/psi*	
	Eucosma hohenwartiana	Knot Grass	*Acronicta rumicis*	
	Eucosma cana	Marbled Beauty	*Cryphia domestica*	
	Lathronympha strigana	Svensson's Copper Underwing	*Amphipyra berbera*	
	Grapholita jungiella	Mouse Moth	*Amphipyra tragopoginis*	
	Grapholita lunulana	Old Lady	*Mormo maura*	
	Cydia ulicetana	Brown Rustic	*Rusina ferruginea*	
	Dichrorampha plumbagana	Small Angle Shades	*Euplexia lucipara*	
	Dichrorampha plumbana	Angle Shades	*Phlogophora meticulosa*	
	Dichrorampha sedatana	Olive	*Ipimorpha subtusa*	
Twenty-plume Moth	*Alucita hexadactyla*	Dingy Shears	*Parastichtis ypsillon*	
Garden Grass-veneer	*Chrysoteuchia culmella*	Dun-bar	*Cosmia trapezina*	
	Crambus lathoniellus	Dark Arches	*Apamea monoglypha*	

Northern Winter Moth, photo Terry Coult

Elephant Hawk Moth, photo Stuart Priestley

Earl Grey Moth, photo Stuart Priestley

45

Water Veneer	Agriphila straminella	Light Arches	Apamea lithoxylaea	
	Agriphila tristella	Clouded-bordered Brindle	Apamea crenata	
	Agriphila inquinatella	Dusky Brocade	Apamea remissa	
	Agriphila latistria	Rustic Shoulder-knot	Apamea sordens	
	Catoptria falsella	Slender Brindle	Apamea scolopacina	
	Acentria ephemerella	Double Lobed	Apamea ophiogramma	
	Scoparia pyralella	Marbled Minor agg.	Oligia strigilis agg.	
	Scoparia ambigualis	Rufous Minor	Oligia versicolor	
	Dipleurina lacustrata	Middle-barred Minor	Oligia fasciuncula	
	Eudonia angustea	Cloaked Minor	Mesoligia furuncula	
	Eudonia mercurella	Rosy Minor	Mesoligia literosa	
Brown China-mark	Elophila nymphaeata	Common Rustic agg.	Mesapamea secalis agg.	
Garden Pebble	Evergestis forficalis	Small Dotted Buff	Photedes minima	

Vapourer Moth, photo Terry Coult

		Small Wainscot	Chortodes pygmina	
	Pyrausta despicata	Flounced Rustic	Luperina testacea	
Small Magpie	Eurrhypara hortulata	Ear Moth agg.	Amphipoea oculea agg.	
	Udea lutealis	Rosy Rustic	Hydraecia micacea	
	Udea prunalis	Frosted Orange	Gortyna flavago	
	Udea olivalis	Bulrush Wainscot	Nonagria typhae	
Rush Veneer	Nomophila noctuella	Uncertain	Hoplodrina alsines	
Mother of Pearl	Pleuroptya ruralis	Rustic	Hoplodrina blanda	
Wax Moth	Galleria mellonella	Mottled Rustic	Caradrina morpheus	
Bee Moth	Aphomia sociella	Pale Mottled Willow	Paradrina clavipalpis	
	Platyptilia gonodactyla	Small Yellow Underwing	Panemeria tenebrata	
	Platyptilia pallidactyla	Green Silver-lines	Pseudoips prasinana	
	Stenoptilia bipunctidactyla	Oak Nycteoline	Nycteola revayana	
	Stenoptilia pterodactyla	Burnished Brass	Diachrysia chrysitis	

Burnished Brass Moth, photo Stuart Priestley

White Plume Moth	Pterophorus pentadactyla	Gold Spot	Plusia festucae	
December Moth	Poecilocampa populi	Lempke's Gold Spot	Plusia putnami gracilis	
Northern Eggar	Lasiocampa quercus f. callunae	Silver Y	Autographa gamma	
Fox Moth	Macrothylacia rubi	Beautiful Golden Y	Autographa pulchrina	
Drinker	Euthrix potatoria	Plain Golden Y	Autographa jota	
Emperor Moth	Saturnia pavonia	Gold Spangle	Autographa bractea	
Pebble Hook-tip	Drepana falcataria	Dark Spectacle	Abrostola triplasia	
Chinese Character	Cilix glaucata	Spectacle	Abrostola tripartita	
Peach Blossom	Thyatira batis	Red Underwing	Catocala nupta	
Buff Arches	Habrosyne pyritoides	Mother Shipton	Callistege mi	
Figure of Eighty	Tethea ocularis	Herald	Scoliopteryx libatrix	
Common Lutestring	Ochropacha duplaris	Small Purple-barred	Phytometra viridaria	
Yellow Horned	Achlya flavicornis	Straw Dot	Rivula sericealis	
Orange Underwing	Archiearis parthenias	Snout	Hypena proboscidalis	
March Moth	Alsophila aescularia	Fan-foot	Zanclognatha tarsipennalis	
Grass Emerald	Pseudoterpna pruinata	Small Fan-foot	Herminia grisealis	

Pebble Hook Tip Moth, photo Stuart Priestley